1 MONTH OF
FREE
READING

at

www.ForgottenBooks.com

By purchasing this book you are eligible for one month membership to ForgottenBooks.com, giving you unlimited access to our entire collection of over 1,000,000 titles via our web site and mobile apps.

To claim your free month visit:

www.forgottenbooks.com/free29817

ISBN 978-1-5279-5082-5
PIBN 10029817

Thornie

DEPARTMENT OF HISTORY OF EDUCATION
ALTHOUSE COLLEGE OF EDUCATION
UNIVERSITY OF WESTERN ONTARIO
LONDON, ONTARIO

AN

ELEMENTARY ARITHMETIC.

BY

G. A. WENTWORTH, A.M.,

AUTHOR OF A SERIES OF TEXT-BOOKS IN MATHEMATICS.

——o○o○——

BOSTON, U.S.A.:

PUBLISHED BY GINN & CO.

PREFACE.

TEACHERS who have the power of putting themselves in the mental attitude of their pupils possess a most important gift. In the first stages of mental growth, as the mind works unseen, it is hard to realize the difficulties encountered, and to decide what assistance can be judiciously given. There is no royal road to the knowledge of arithmetic, but the steps can be made short and easy. The little learner need not be wearied, if the exercise is not too long continued. He may also have the consciousness of effort, as in learning to walk, and above all, the pleasure of succeeding. This result can be secured only by observing the following fundamental principles:

1. **All elementary teaching of arithmetic must be begun by the pupils observing and handling objects.**

It is surprising that school authorities in many places decline to furnish the money, however small the amount is, to purchase the simple apparatus required for each Primary School. They assert that they got on well enough without such aids when they were children, and they seem to be quite unconscious of the weakness of such an argument. The question is not whether children can do well without the aid of objects, but whether they can do better with them. Our fathers did well enough in travelling on horseback and in coaches, but we do better with our express trains. Children may be able to grasp abstract ideas, after sufficient time, without the aid of concrete examples. It is certain, however, that they grasp these ideas more firmly and more quickly if they are led to them by easy steps through objects that can be seen and handled.

In the treatment of each number we must rely upon the *sight* of the pupil, and not upon his *hearing*. Furthermore, we must rely upon his *activity*. He must *do* as well as *see*. Listless repetition of 4 and 3 are 7, or the sing-song 4 times 3 are 12, makes no impression upon him. The next day he is quite likely to tell you that 4 and 3 are 6. If he is required to put 4 pegs in one row of the counting-board and 3 in another row, and to learn in this way that 4 and 3 are 7, he will remember it. This method of teaching has the very great advantage of giving to the study of arithmetic the peculiar distinction that the learner can discover for himself, in case of doubt, whether his answer to any question is right or wrong, and can find the true answer, if he has given a wrong one.

2. A knowledge of the processes of arithmetic should be acquired by using small numbers; and each number should be treated in all its variations before the next higher number is considered.

By means of a small number of objects children can easily acquire clear notions of the different processes of arithmetic, and at the outset begin to form the habit of being *independent of rules*. With 6 blocks they can learn to add 4 and 2, to subtract 4 from 6, to multiply 3 by 2, to divide 6 by 2, without suspecting the existence of the fearful rules to be found in our text-books of arithmetic. They can also be taught to find $\frac{1}{2}$ of 6 or $\frac{2}{3}$ of 6, without even hearing of the terms, fraction, numerator or denominator.

Variety is indispensable for securing and maintaining the pupils' interest in number work. This is a sufficient reason for treating each number in all possible ways before taking up the next higher number. Besides, this method is the natural method, and has stood the test of experience. It is found to be the best method for giving practical mastery of the facts of each number; in short, it secures the greatest interest and the best results.

3. **Repetition is to be regular and systematic, combined with suitable variation.**

It cannot be too strongly urged that the first requisite of good teaching is repetition, the second requisite is repetition, and the third requisite is repetition. The interest of the pupil must be kept up *by varying the application of the question.* To find the sum of 3 horses and 5 horses is not the same thing to the child as to find the sum of 3 tops and 5 tops. Hence a lesson may be given as many times as may be necessary by properly varying the questions.

A table of different things, given opposite the first page of this book, will be found of great use in suggesting a suitable variety of questions. Care must be exercised to have the variation of a kind to *fix* knowledge. To ask the number of 3 ducks and 4 ducks, of 3 times 4 ducks, and $\frac{1}{3}$ of 12 ducks, *in succession,* is a variation, to be sure, but of a kind to distract the child's mind, as he cannot quickly pass from one conception to the other. The questions in Part I. of this book are specimen questions, which it is expected the Teacher will supplement by a great number and variety of other questions.

4. **Lessons should be short, answers required should be simple, and the power to deal with numbers in the abstract should be acquired through concrete examples by regular gradation.**

Number work should be discontinued the moment the pupil's attention flags. It is far better to divide the time daily allotted to arithmetic into two or more lessons. Only simple, direct answers should be required. Of course, if objects are named in the question, they should be named in the answer. The answer to 5 birds + 3 birds should be 8 *birds*, and not simply 8.

A knowledge of numbers in the abstract is obtained only by a comparison of different things. The child learns the number 5, for instance, by seeing and handling 5 familiar objects, by observing number pictures of 5 on the blackboard or on cardboard, by answering questions about 5 familiar but unseen objects, and lastly about 5 in the abstract.

5. **The child must not be required to read questions that are difficult for him to read, or to solve problems that are difficult for him to analyze.**

The intention is to put this book into the hands of young pupils, *but only for them to copy and do the numerical exercises.* The other examples, usually called *clothed* examples by way of distinction, must be read by the Teacher, and *only* the answers be required of pupils. No child can become interested or successful in arithmetic if his mind is *distracted* between the *reading* of a problem and the *numerical calculation* required for its solution. He can learn the simple processes of arithmetic while quite young; he can learn to be accurate and reasonably rapid in these processes; he can learn to be neat and orderly in the arrangement of his work; and his interest will constantly increase, *provided he is kept master of his field of operations.* At this early stage he cannot be exercised in logical analysis, and it is a great mistake to put problems before him that require too great an exercise of the reasoning faculty. Later he will form the habit of close attention, learn the meaning of logical inference, and acquire the power of sustained and continuous thought.   Arithmetic rightly taught furnishes the very essence of intellectual training, and deserves the name of "The Logic of the People."

These principles are recognized and adopted by the best educators everywhere, but for the concise statement of them I am in some measure indebted to a distinguished school-inspector of Australia.

I have to thank many eminent Teachers for valuable suggestions and criticisms.   I shall be grateful for further suggestions, and especially for any corrections.

G. A. WENTWORTH.

Exeter, N.H., 1893.

TABLE FOR VARYING QUESTIONS.

Animals ... Dog, Puppy, Cat, Kitten, Rabbit, Cow, Calf, Pig, Horse, Colt, Sheep, Lamb, Goat, Kid, Fox, Mouse, Squirrel, Monkey.

Birds ...... Robin, Sparrow, Swallow, Canary, Parrot, Crow, Bluebird, Kingbird, Hawk, Owl, Jay, Loon, Swan, Pigeon.

Clothes .... Hat, Cap, Bonnet, Coat, Vest, Dress, Socks, Boots, Shoes, Collar, Cuffs, Slippers, Rubbers, Mittens, Gloves.

Flowers .... Rose, Pink, Daisy, Pansy, Lily, Geranium, Violet, Poppy.

Fowls ..... Hen, Chicken, Turkey, Duck, Goose, Gosling.

Fruits ..... Apple, Pear, Quince, Orange, Lemon, Peach, Grape, Fig.

Garden ..... Peas, Beans, Corn, Potatoes, Carrots, Parsnips.

House ..... Room, Door, Window, Chair, Table, Picture, Carpet, Cup, Plate, Saucer, Fork, Knife, Spoon, Pitcher, Clock.

Insects ..... Fly, Spider, Bee, Hornet, Butterfly, Beetle, Cricket.

School ..... Desk, Slate, Pencil, Pen, Book, Paper, Chair.

Smallwares .. Buttons, Pins, Needles, Spools of Thread.

Store ...... Tea, Coffee, Sugar, Starch, Soap, Candles, Matches, Eggs, Axe, Rake, Pail, Spade, Hoe, Saw, Nails.

Toy-Store ... Doll, Top, Ball, Whip, Basket, Marbles, Whistle.

Tradesmen .. Baker, Butcher, Grocer, Milkman, Blacksmith.

Trees ...... Apple, Oak, Cherry, Plum, Ash, Birch, Beech.

Vehicles ... Train, Car, Coach, Hack, Buggy, Wagon, Gig, Sleigh, Sled, Barge, Bus.

ELEMENTARY ARITHMETIC.

Part I.

Part I. is intended as a guide to teachers in oral and blackboard work for children before they can read. After they can read, a rapid review will help fix their knowledge of simple arithmetical processes.

THINGS NEEDED.

1. **Objects for Counters.** Such as cents, blocks, buttons, spools, pencils, nails, little tin plates, cups and saucers, inch-squares of pasteboard, foot-rules, yard-sticks, a set of tin measures for liquids, a set of wooden measures for dry articles, and a set of weights.

2. **A Counting-Board.** This is of great assistance in teaching arithmetical processes with small numbers. It is simply a smooth board with 100 holes about an inch apart, arranged in 10 rows of 10 holes each. Nails or wooden pins can be used for counters.

Another way of making the counting-board is to drive 100 nails in 10 rows of 10 nails each through a piece of board, at suitable distances from each other, until they project about an inch, and use spools for counters, slipping them on the ends of the nails.

LESSON 1.

THE NUMBER ONE.

Show me *one* finger; *one* block; *one* button.

How many suns do we see by day? How many moons by night?

We write the **figure 1** for **one**.

NOTE. The introduction of figures may be postponed until after the number six is taught. In that case some variation in the language will be required.

THE NUMBER TWO.

How many fingers are *one* finger and *one* finger?

Hold up **two** fingers; **two** hands.

We write the **figure 2** for **two.**

NOTE. Pictures of balls, cups, tops, blocks, etc., are introduced in places where it is expected the Teacher will show objects of some kind.

How many balls are ◎ and ◎?

How many cups are ⊔ and ⊔?

How many dolls are 1 doll and 1 doll?

How many horses are 1 horse and 1 horse?

How many are 1 and 1?

Here are two blocks, ▨ ▨. Take away ▨. How many are left?

1 apple from 2 apples leaves how many?

1 from 2 leaves how many?

How many more pears are △ △ than △?

How many more dolls are 2 dolls than 1 doll?

How many rings must you put with ○ to have ○ ○?

How many apples must you put with 1 apple to have 2 apples?

NOTE. The following plan is recommended to the Teacher, for the number-work of Part I.:

1. Show objects, and secure the desired result from them.

2. Draw pictures of blocks, squares, etc., on the board, and obtain the same result from the pictures.

3. Ask the same question on familiar but unseen objects.

4. Finish with abstract numbers.

The Teacher can vary the questions at pleasure by using different objects and different pictures, and by using the table of familiar objects given opposite the first page.

THE NUMBER THREE.

How many fingers are *two* fingers and *one* finger?

Hold up **three** fingers.

We write the **figure 3** for three.

* Copy each card below, and write under it the figure for the number of dots in the card:

Count the dots in these cards from left to right.

Count the dots from right to left.

What number follows 1? What number follows 2?

What number comes before 2? before 3?

What number is between 1 and 3?

* Copy these pictures, and write under each group the figure for the number in the group.

△ △ △   □ □ □   ● ● ●

× × ×   ○ ○ ○   * * *

How many pears are ⌀ and ⌀ and ⌀?

How many balls are ◎ and ◎ and ◎?

How many dogs are 1 dog and 1 dog and 1 dog?

How many boys are 1 boy and 1 boy and 1 boy?

How many are 1 and 1 and 1?

How many stars are * * and *?

* NOTE. In this case and in similar cases the Teacher should put the number pictures on the board, and then require the pupils to follow the directions given. The Teacher should require the attention of the pupils only a few minutes at a time. One of these "Lessons" will make a great many lessons for the children.

How many apples are 🍎 and 🍎 🍎 ?

How many pinks are 1 pink and 2 pinks ?

How many are 1 and 2 ? 2 and 1 ?

Here are three blocks, ▨ ▨ ▨

Take 1 block away, how many will be left ?

Take 2 blocks away, how many will be left ?

Take 3 blocks away, how many will be left ?

How many more blocks are ▨ ▨ ▨ than ▨ ▨ ?

How many more cows are 3 cows than 2 cows ?

How many more figs are 3 figs than 1 fig ?

How many blocks must be put with ▨ to make ▨ ▨ ▨ ?

How many baskets must be put with 🧺 🧺 to make 🧺 🧺 🧺 ?

How many plums must be put with 2 plums to make 3 plums ?

How many plums must be put with 1 plum to make 3 plums ?

James may take 1 block; then 1 more; and then 1 more.

How many *times* has James taken 1 block ? How many blocks has he ? Then 3 times 1 block are how many blocks ?

How many chairs are 3 times 1 chair ?

Here are 3 apples, 🍎 🍎 🍎. How many boys can each have 1 apple ?

Here are 3 dolls. How many girls can each have 1 doll ? How many *ones* in 3 ?

THE NUMBER FOUR.

Three dots and *one* dot make **four dots.**
Here are *four* dots, ● ● ● ●
We write the **figure 4** for **four.**

Copy each card below, and write under it the figure for the number of dots in the card:

Count the dots in these cards from left to right.
Count the dots from right to left.
What number follows 2? What number follows 3?
What number comes before 4? before 3?
What number is between 1 and 3? 2 and 4?

Copy these pictures, and write under each group the figure for the number in the group:

✳ ✳ ✳ ✳   ○ ○ ○ ○

✚ ✚ ✚ ✚   ☐ ☐ ☐ ☐

How many sides has this square ☐ ?
How many legs has a horse? a frog? a cow?
How many stars are ✳ ✳ ✳ and ✳ ?
How many rings are ○ ○ ○ and ○ ?
How many crosses are + and + + + ?
How many eggs are ○ and ○ ○ ○ ?
How many boys are 3 boys and 1 boy?
How many mice are 1 mouse and 3 mice?
How many are 3 and 1? 1 and 3?

How many stars are ✳ ✳ and ✳ ✳ ?
How many marks are // and // ?
How many brooms are 2 brooms and 2 brooms ?
How many are 2 and 2 ?

Here are four blocks, ▣ ▣ ▣ ▣
Cover one block. How many can you see ?
Then 1 from 4 leaves how many ?
Cover two blocks. How many can you see ?
Then 2 from 4 leaves how many ?
Cover three blocks. How many can you see ?
Then 3 from 4 leaves how many ?
Cover all four blocks. How many can you see ?
Then 4 from 4 leaves how many ?

How many more tops are 🌀🌀🌀🌀 than 🌀🌀🌀 ?
How many more balls are ◎◎◎◎ than ◎◎ ?
How many more crosses are ✠✠✠✠ than ✠ ?
 * How many more cars are 4 cars than 3 cars ?
than 2 cars ? than 1 car ?

How many more apples are 4 apples than 2
apples ? than 1 apple ? than 3 apples ?

How many ladders must be put with 目 to make
目 目 目 目 ?
How many pears must be put with ◁ ◁ to
make ◁ ◁ ◁ ◁ ?
How many crosses must be put with **X X X** to
make **X X X X** ?

* The Teacher must make the question complete in each case.
Thus, How many more cars are 4 cars than 3 cars ? How many more
cars are 4 cars than 2 cars ? How many more cars are 4 cars than 1 car ?

Here are 4 blocks, ▢ ▢ ▢ ▢
Susie may take 1 block; then 1 more; then 1 more; and then 1 more. How many *times* has Susie taken 1 block? How many blocks has she? Then how many blocks are 4 times 1 block?

How many apples are 4 times 1 apple?
How many figs are 4 times 1 fig?
How many are 4 times 1?

Ernest may take 2 blocks; and then 2 more. How many *times* has Ernest taken 2 blocks? How many blocks has he? Then how many blocks are 2 times 2 blocks?

How many cakes are 2 times 2 cakes?
How many rolls are 2 times 2 rolls?
How many are 2 times 2?

Here are 4 apples, ◌ ◌ ◌ ◌. How many boys can have 1 apple each? How many *ones* in 4?

Here are 4 tops, ◌ ◌ ◌ ◌. How many boys can have 2 tops each? How many *twos* in 4?

Here are 4 dots, ○ ● ● ○
Divide them into *two equal parts*, thus ● ●/● ○
How many dots in each part?

When a number of things is divided into **two equal parts,** each part is called **one-half** of the whole number.

What is one-half of

4 blocks?	4 pears?	4 cents?
4 books?	4 buns?	2 oranges?

THE HALF OF A UNIT.

If an apple is cut into two equal parts, what is one of the parts called?

What are the two parts called?

If an orange is divided into two equal parts, what is one of the parts called?

If anything is divided into two equal parts, what is one of the parts called?

How many halves of an apple in a whole apple?

How many halves of an orange make the whole orange?

How many halves of anything make the whole thing?

How many times must you take one-half of an apple to have an apple?

If one-half of an orange is worth 2 cents, how many cents is the orange worth?

If an apple is worth 2 cents, how much is one-half of the apple worth?

How many halves of an apple in two whole apples?

How many halves of an apple in one whole apple and one-half of another apple?

Draw a straight line and divide it into halves.

Draw a square and divide it into halves.

THE NUMBER FIVE.

Four dots and *one* dot make five dots.

Here are *five* dots, °₀°

We write the **figure 5** for **five.**

How many fingers on your right hand?

How many fingers on your left hand?

Copy these pictures, and write under each group the figure for the number in the group:

Copy each card below, and write under it **the figure** for the number of dots in the card:

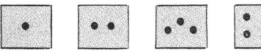

Count **the** dots in these cards from left to right.

Count the dots from right **to** left.

What number follows 4? What number follows 2?

What number comes before 5? before 2? before 4?

What number is between 3 and 5? 2 and 4?

How many stars are ✳ ✳ ✳ ✳ and ✳?

How many tops are � ♧ ♧ ♧ and ♧?

How many ladders are ⧉ and ⧉ ⧉ ⧉ ⧉?

How many crosses are + and + + + +?

How many apples are 4 apples and 1 apple?

How many plums are 1 plum and 4 plums?

How many are 4 and 1? 1 and 4?

How many marks are / / and / / / ?

How many bottles are 🍶 🍶 and 🍶 🍶 🍶 ?

How many balls are ◉ ◉ ◉ and ◉ ◉ ?

How many mugs are ♨ ♨ ♨ and ♨ ♨ ?

How many figs are 3 figs and 2 figs ?

How many spoons are 2 spoons and 3 spoons ?

How many cars are 3 cars and 2 cars ?

How many lambs are 3 lambs and 2 lambs ?

How many are 2 and 3 ?

How many are 3 and 2 ?

Here are five blocks, ▢ ▢ ▢ ▢ ▢

Cover one block.   How many can you see ?

Then 1 from 5 leaves how many ?

Cover two blocks.   How many can you see ?

Then 2 from 5 leaves how many ?

Cover three blocks.   How many can you see ?

Then 3 from 5 leaves how many ?

Cover four blocks.   How many can you see ?

Then 4 from 5 leaves how many ?

Cover five blocks.   How many can you see ?

Then 5 from 5 leaves how many ?

How many more dots are ● ● ● ○ ○ than ● ● ● ● ?

How many more stars are ✳ ✳ ✳ ✳ than ✳ ✳ ✳ ?

How many more crosses are × × × × than × × ?

How many more balls are ◉ ◉ ◉ ◉ ◉ than ◉ ?

How many more hens are 5 hens than 3 hens ?

How many more dogs are 5 dogs than 2 dogs ?

How many more lambs are 5 lambs than 4 lambs?

How many more pigs are 5 pigs than 1 pig ?

How many rings must be put with ◯ to make
◯ ◯ ◯ ◯ ◯ ?

How many stars must be put with ✳ ✳ ✳ to
make ✳ ✳ ✳ ✳ ✳ ?

How many crosses must be put with × × to
make × × × × × ?

How many marks must be put with /// to make
///// ?

How many cents must be put with 2 cents to
make 5 cents ?

How many balls must be put with 3 balls to
make 5 balls ?

How many pigeons must be put with 1 pigeon
to make 5 pigeons ?

How many pears must be put with 4 pears to
make 5 pears ?

If Frank takes 1 block at a time for 5 times,
how many blocks will he have ?

Then 5 times 1 block are how many blocks ?

5 times 1 apple are how many apples ?

5 times 1 cup are how many cups ?

Here are 5 tops, ♉ ♉ ♉ ♉ ♉

How many boys can have 1 top apiece ?

How many *ones* in 5 ?

How many boys can have 2 tops apiece ?

How many will be left for another boy ?

How many *twos* in 5, and how many over ?

How many *threes* in 5, and how many over ?

How many *fours* in 5, and how many over ?

THE NUMBER SIX.

Five dots and *one* dot make **six** dots.

Here are *six* dots, ⦂•

We write the **figure 6** for **six**.

Copy these pictures, and write under each group the figure for the number in the group :

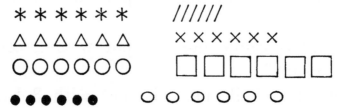

Copy each card below, and write under it the figure for the number of dots in the card :

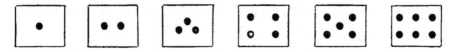

Count these dots from left to right.

Count these dots from right to left.

What number follows 4? What number follows 5?

What number comes before 4? before 6? before 5? before 3? before 2?

What number is between 4 and 6? 3 and 5?

How many balls are ◎ ◎ ◎ ◎ ◎ and ◎?

How many tops are ♉ and ♉ ♉ ♉ ♉ ♉?

How many boxes are 5 boxes and 1 box?

How many brooms are 1 broom and 5 brooms?

How many birds are 5 birds and 1 bird?

How many oranges are 1 orange and 5 oranges?

How many are 5 and 1? 1 and 5?

How many stars are ✳ ✳ ✳ ✳ and ✳ ✳ ?

How many rings are ◯ ◯ and ◯ ◯ ◯ ◯ ?

How many balls are ◎ ◎ and ◎ ◎ and ◎ ◎ ?

How many kittens are 4 kittens and 2 kittens?

How many horses are 4 horses and 2 horses?

How many buns are 2 buns and 4 buns?

How many pies are 2 pies and 4 pies?

How many are 4 and 2? 2 and 4?

How many crosses are ✠ ✠ ✠ and ✠ ✠ ✠ ?

How many apples are 3 apples and 3 apples?

How many are 3 and 3 ?

Here are 6 blocks, ▱ ▱ ▱ ▱ ▱ ▱

Cover 1 block.   How many can you see?

Then 1 from 6 leaves how many?

Cover 2 blocks.   How many can you see?

Then 2 from 6 leaves how many?

Cover 3 blocks.   How many can you see?

Then 3 from 6 leaves how many?

Cover 4 blocks.   How many can you see?

Then 4 from 6 leaves how many?

Cover 5 blocks.   How many can you see?

Then 5 from 6 leaves how many?

Cover 6 blocks.   How many can you see?

Then 6 from 6 leaves how many?

How many more dots are ●●●●●● than ●●●●● ?

How many more stars are ✳✳✳✳✳✳ than ✳✳✳✳ ?

How many more crosses are ×××××× than ××× ?

How many more marks are ////// than // ?

How many more tops are ♥♥♥♥♥♥ than ♥ ?

How many more chairs are 6 chairs than 5 chairs?

How many more boxes are 6 boxes than 4 boxes?

How many more cars are 6 cars than 3 cars?

How many more dogs are 6 dogs than 2 dogs?

How many more pears are 6 pears than 1 pear?

How many marks must be put with ///// to make //////?

How many tops must be put with 🌂 🌂 🌂 🌂 to make 🌂 🌂 🌂 🌂 🌂 🌂?

How many stars must be put with * * * to make * * * * *?

How many balls must be put with ◎ ◎ to make ◎ ◎ ◎ ◎ ◎ ◎?

How many squares must be put with □ to make □ □ □ □ □ □?

How many bells must be put with 3 bells to make 6 bells?

How many caps must be put with 2 caps to make 6 caps?

How many pies must be put with 4 pies to make 6 pies?

How many cups must be put with 1 cup to make 6 cups?

How many books must be put with 5 books to make 6 books?

Here are 6 blocks, ▨ ▨ ▨ ▨ ▨ ▨

If Hattie takes 2 blocks at a time for 3 times, how many blocks will she have?

Then how many blocks are 3 times 2 blocks?

Here are 6 blocks, ▨▨▨ ▨▨▨
John may take 3 blocks; then 3 more.
How many times has John taken 3 blocks?
How many blocks has he?
Then how many blocks are 2 times 3 blocks?
How many oranges are 2 times 3 oranges?
How many are 2 times 3?

Here are 6 apples, ◌ ◌ ◌ ◌ ◌ ◌
How many girls can have 1 apple each?
How many *ones* in 6?
How many girls can have 2 apples each?
How many *twos* in 6?
How many girls can have 3 apples each?
How many *threes* in 6?

Here are 6 dots, ● ● ● ● ● ●
Divide them into two equal parts, ●●●/●●●
How many dots are there in each part?
What is one-half of 6 dots? 6 cents?
What is one-half of 4 apples? 2 pens?
Divide 6 dots into *three* equal parts, thus,
●●/●●/●●
How many dots are there in each part?
When a number of things is divided into **three
equal parts**, each part is **one-third** of the number.
What is one-third of 6 dots? of 6 cents?
How many dots are two-thirds of 6 dots?
How many dots are three-thirds of 6 dots?
How many oranges are two-thirds of 6 oranges?

Into how many equal parts is this pineapple cut?

What is one of the parts called?

What are two of the parts called?

What are the three parts together called?

If a circle is cut into three equal parts, what is one of the parts called? two of the parts? the three parts?

How many thirds of an apple in one apple?

How many thirds of an apple in two apples?

How many thirds of an apple in one apple and one-third of an apple?

How many thirds of an apple in one apple and two-thirds of an apple?

If an orange is worth 3 cents, how many cents is one-third of the orange worth?

If one-third of a big stick of candy is worth 2 cents, how many cents is the whole stick worth?

John had a stick of candy, but he gave his little sister one-third of it. How many thirds of the stick had he left?

James had to walk from his house to the schoolhouse. After he had walked two-thirds of the way, how many more thirds had he to walk?

How many apples at 2 cents apiece can you buy for 4 cents?

NOTE. The answer required should be simply : **2 apples.**

How many boots does it take to make a pair of boots? How many horses to make a pair of horses?

How many pairs of boots does it take for 3 boys?

How many boots in 2 pairs of boots? How many horses in 3 pairs of horses? How many oxen in 3 pairs of oxen?

The butcher has 2 horses, the grocer 2 horses, and the baker has 1 horse. How many horses have they in all?

Mary has 3 cages, and 1 bird in each cage. How many birds has she?

There were 5 sheep in the pasture, and each sheep had 1 lamb. How many lambs were there?

There were 5 apples on a limb. Three fell off. How many were left?

Harold had 5 cents, and bought a ball for 2 cents. How many cents did he have then?

A blacksmith had 6 horses to shoe. He shod half of them. How many more had he to shoe?

A blacksmith shod 4 horses before dinner, and 2 after dinner. How many did he shoe?

John and his papa hoed 6 rows of corn. John hoed one-third of the 6 rows, and his papa two-thirds of them. How many rows did each hoe?

What part of 6 apples are 2 apples?

What part of 6 apples are 3 apples?

At 3 cents apiece how many oranges can you buy for 6 cents?

At 2 cents apiece how many apples can you buy for 6 cents?

If you can buy 3 sticks of candy for 3 cents, how many sticks can you buy for 4 cents?

If you have 6 eggs, 2 on a plate, how many plates have you?

If you can buy 2 apples for 2 cents, how many apples can you buy for 6 cents?

What is one-half of 6 cents? one-third of 6 cents? two-thirds of 6 cents? three-thirds of 6 cents?

Charlie sold 3 newspapers for 2 cents a paper. How many cents did he get for the 3 papers?

It takes 2 cents to buy a paper. How many papers can you buy for 4 cents? for 6 cents?

If you have six cents, and spend half of your money, how many cents will you have left?

How many balls in one-half of 6 balls? in two-halves of 6 balls?

How many blocks in one-third of 6 blocks? in two-thirds of 6 blocks? in three-thirds of 6 blocks? in three-thirds of 3 blocks?

If a cook has 6 eggs, and uses one-third of them for cake, how many eggs will be left?

A little boy had 4 newspapers to sell, and he sold half of them. How many papers had he left?

How many pears are one-half of 4 pears and one-half of 6 pears together?

THE NUMBER SEVEN.

Six dots and *one* dot make **seven** dots.

Here are seven dots, ●●●●

We write the **figure** 7 for **seven**.

Draw these cards, and write 7 under each card.

How many are 3 and 4? 4 and 3? 1 and 6? 2 and 5? 5 and 2? 6 and 1?

How many are 7 less 3? 7 less 4? 7 less 1? 7 less 2? 7 less 5? 7 less 6? 7 less 7?

How many more are 7 chairs than 3 chairs? 7 balls than 4 balls? 7 kittens than 1 kitten? 7 mice than 5 mice? 7 ladders than 2 ladders? 7 stars than 6 stars?

How many dolls must you put with 3 dolls to have 7 dolls?

How many cups must you put with 4 cups to have 7 cups?

How many hats must you put with 1 hat to have 7 hats?

How many cents must you put with 2 cents to have 7 cents?

How many eggs must you put with 5 eggs to have 7 eggs?

How many cents must you put with 6 cents to have 7 cents?

There are 4 pigs in 1 pen and 3 pigs in another pen. How many pigs in both pens?

How many must you add to 5 to make 7?

If you draw 7 stars and rub out 3 of them, how many will be left? How many are 7 less 3?

How many crosses are 6 crosses and 1 cross?

If you have 7 pears and give away 6 of them, how many pears will you have left?

How many are 7 less 6? 7 less 1? 7 less 3? 7 less 5? 7 less 2? 7 less 4?

How many mittens make a *pair* of mittens?

How many boots make a *pair* of boots?

Here are 7 blocks, 🎲🎲🎲 🎲🎲🎲🎲. Call them horses, and find how many pairs of horses you can have, and how many single horses besides?

NOTE. Show the pupils one-cent, two-cent, and five-cent coins. Let them count out the number of single cents a two-cent piece equals in value, and the number a five-cent piece equals in value. Show the one-cent, two-cent, three-cent, four-cent, and five-cent postage stamps.

At 1 cent apiece, how many apples can you buy for a two-cent piece? for a five-cent piece?

Harry has a two-cent piece and a five-cent piece. How many one-cent postage stamps can he buy?

If you add 1 block to 3 times 2 blocks, how many blocks will you have?

How many are □ □ and □ □ and □ □ and □?

How many are □ □ □ and □ □ □ and □?

How many are 2 and 2 and 2 and 1?

How many are 3 and 3 and 1?

Alice has a five-cent piece and a two-cent piece, and Harry has six cents. How much more money has Alice than Harry?

How many peaches are 3 peaches and 4 peaches?

A farmer had 7 horses. If he had 3 turned out in the pasture, and the rest in the stable, how many did he have in the stable?

There were 7 windows in a room, and 2 of them were shut. How many were open?

There were 7 eggs in a basket, but the cook used 5 of them. How many were left?

A storekeeper had 7 saws. He sold one saw to a carpenter. How many had he left?

A man had 7 cows to milk. When he had milked 6 cows, how many had he to milk?

If one apple costs 1 cent, how much will 7 apples cost? 5 apples? 3 apples? 6 apples?

If one peach costs 2 cents, how much will 3 peaches cost? 2 peaches?

If you can buy one pencil for 2 cents, how many pencils can you buy for 6 cents? for 4 cents?

How many three-cent stamps can you buy for 3 cents? for 6 cents?

George has 7 cents. How many oranges can he buy at 3 cents each, and how many cents will he have left?

Ellen has 7 cents. How many pears can she buy at 2 cents each, and how many cents will she have left?

DRILL EXERCISE.

NOTE. The Teacher may put the following groups of dots on the board, and call upon the pupils *one by one* to tell the number of dots as she touches the squares at random, with a pointer. *Every child* should be carefully drilled on this exercise until he can name each number of dots instantly.

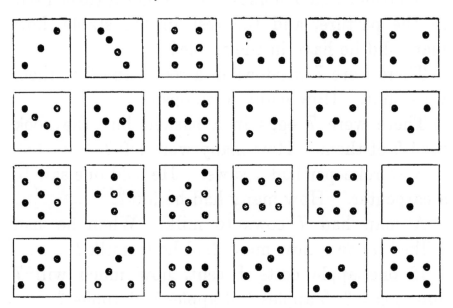

Name two numbers that together make 4.

Name two numbers that together make 5.

Name three numbers that together make 5.

Name two numbers that together make 6.

Name three numbers that together make 6,

Name two numbers that together make 7.

Name three numbers that together make 7·

Name four numbers that together make 7,

How many more are 6 than 4 ? than 3 ?

How many more are 5 than 3 ? than 2 ?

How many more are 7 than 4 ? than 5 ?

How many more are 7 than 3 ? than 2 ?

THE SIGNS = AND +.

The sign = stands for the word **are** or **is.**
Copy, and use the sign that stands for *are :*

1 and 1	2.	5 and 1	6.
2 and 1	3.	2 and 5	7.
4 and 3	7.	3 and 2	5.
2 and 4	6.	2 and 2	4.

The sign + stands for the word **and.**
Copy, and use the sign that stands for *and :*

3	1 = 4.	3	3 = 6.
1	2 = 3.	3	4 = 7.
4	2 = 6.	5	2 = 7.
2	3 = 5·	1	5 = 6.

Copy, and write each answer at the right of the sign = :

1 + 1 =	1 + 2 =	3 + 1 =
2 + 4 =	1 + 4 =	4 + 1 =
3 + 2 =	3 + 3 =	1 + 5 =
1 + 3 =	5 + 2 =	3 + 4 =
2 + 2 =	4 + 2 =	6 + 1 =
2 + 1 =	1 + 6 =	4 + 3 =

1 + 1 + 1 =	2 + 2 + 2 =
2 + 1 + 1 =	1 + 2 + 3 =
2 + 2 + 1 =	2 + 3 + 2 =
3 + 1 + 2 =	3 + 3 + 1 =
3 + 2 + 2 =	2 + 3 + 1 =
5 + 1 + 1 =	4 + 1 + 2 =

THE SIGNS — AND ✕.

The sign — stands for the word **minus**.

When we take 3 blocks from 5 blocks, we have 2 blocks left.

We write this,

$$5 \text{ blocks} - 3 \text{ blocks} = 2 \text{ blocks}:$$

and we read this,

$$5 \text{ blocks } \textit{minus } 3 \text{ blocks are } 2 \text{ blocks.}$$

Oral and slate exercises :

BLOCKS.	BLOCKS.	BLOCKS.
$3 - 1 =$	$6 - 1 =$	$6 - 4 =$
$2 - 1 =$	$6 - 3 =$	$5 - 4 =$
$4 - 2 =$	$7 - 1 =$	$7 - 5 =$
$5 - 3 =$	$5 - 2 =$	$7 - 3 =$
$3 - 2 =$	$7 - 2 =$	$7 - 4 =$

The sign ✕ stands for the word **times.**

PEARS.	PEARS.	PEARS.
$3 \times 1 =$	$2 \times 3 =$	$5 \times 1 =$
$2 \times 1 =$	$4 \times 1 =$	$7 \times 1 =$
$2 \times 2 =$	$6 \times 1 =$	$3 \times 2 =$

NOTE. The pupils should copy the above, and similar exercises, on blocks of paper or slates, and write the answer for each example.

Also the Teacher should put these exercises on the blackboard, and with pointer in hand require of *each pupil in turn* quick answers to such examples as she touches with the pointer. One child at a time should give the answers aloud, and the other members of the class should be on the alert to raise their hands when a wrong answer is given. If a child gives a wrong answer, he should be sent to the counting-board to discover the true answer.

THE DAYS OF THE WEEK.

On what day of the week do we go to church?

The next day we come to school. Who can tell the name of the day that follows Sunday?

Who can tell the name of the day that follows Monday?

What day comes after Tuesday?

What day comes after Wednesday?

What day comes after Thursday?

The day that follows Friday we have for a holiday. What is the name of that day?

We will write the first letter of each day, thus:

S ——————,
M——————,
T ——————,
W——————,
T ——————,
F ——————,
S ——————.

Repeat with me the days of the week, beginning with Sunday.

Wednesday is the middle day of the week.

What day comes before Wednesday?

How many days make a week?

Remember : **7 days make 1 week.**

Note. The Teacher should make sure that *all* the pupils of the class give close attention and learn the days of the week, and not be satisfied if *some one* in the class can repeat them. In fact, *this caution applies to all class-work.*

THE NUMBER EIGHT.

Seven dots and *one* dot make **eight** dots.

Here are *eight* dots, ● ● ● ●

We write the **figure 8** for **eight.**

Copy these pictures and write under each group the figure for the number in the group :

××××××××   ✱✱✱✱✱✱✱✱

Copy each card below, and write under it the figure for the number of dots in the card.

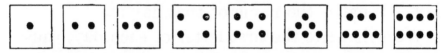

Count the dots in these cards from left to right.

Count the dots from right to left.

What number follows 5 ? follows 7 ?

What number comes before 8 ? comes before 5 ?

What number is between 5 and 7 ? 6 and 8 ?

Copy, and add dots enough to make 8 dots in each card below :

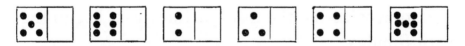

How many blocks are 5 and 3 ? 6 and 2 ? 4 and 3 ? 4 and 4 ? 2 and 5 ? 2 and 6 ? 7 and 1 ?

How many dots must be put with 5 to make 8 ? with 2 to make 8 ? with 3 to make 8 ? with 4 to make 8 ? with 7 to make 8 ? with 6 to make 8 ?

How many more dots are 8 than 6 ? 8 than 3 ? 8 than 4 ? 8 than 2 ? 8 than 1 ? 8 than 5 ?

Here are 8 blocks, ▨▨ ▨▨ ▨▨ ▨▨

If you take away 2 blocks, how many will be left?

If you take away 6 blocks, how many will be left?

If you take away 5 blocks, how many will be left?

If you take away 3 blocks, how many will be left?

If you take away 4 blocks, how many will be left?

If you take away 1 block, how many will be left?

If you take away 7 blocks, how many will be left?

Ellen may take 2 blocks at a time for 4 times. How many blocks has she? How many blocks, then, are 4 times 2 blocks?

How many cups are 4 times 2 cups?

How many pears are 4 times 2 pears?

Erwin may take 4 blocks, and then 4 more. How many times has he taken 4 blocks? How many blocks has he? How many blocks, then, are 2 times 4 blocks?

How many plums are 2 times 4 plums?

How many apples are 2 times 4 apples?

How many are 4 times 2? How many are 2 times 4? How many 2's in 8? How many 4's in 8?

How many are 3 times 2? How many are 2 times 3? How many 2's in 6? How many 3's in 6?

How many oranges are one-half of 6 oranges?

How many apples are one-third of 6 apples?

When we take *one-half* of 6 oranges, into how many *equal parts* do we divide the 6 oranges?

When we take *one-third* of 6 apples, into how many *equal parts* do we divide the 6 apples?

Here are 8 blocks,

How many times must Nora go to bring these blocks to me if she brings just 2 blocks each time? Then 8 blocks divided by 2 blocks = 4 *times*.

But if Nora divides the blocks into two equal parts, how many blocks will there be in each part? Then 8 blocks divided by 2 = 4 *blocks*.

NOTE. The Teacher must illustrate in many ways the two different meanings of Division. When the divisor is a mere number, as 2, 3, 4, etc., the meaning of division then is the separation of the given number of things into 2, 3, 4, etc., equal parts, and the quotient will signify *a number of things like the dividend.* When the divisor is a *number of things like the dividend*, the quotient will signify *the number of times* the divisor is contained in the dividend; that is, the number of times the divisor can be taken from the dividend.

How many times are 2 cents contained in 6 cents?
How many times are 2 cents contained in 8 cents?
How many times are 4 cents contained in 8 cents?
How many times are 3 cents contained in 6 cents?

What is the answer for

    8 cents divided by 4 cents?
    8 pears divided by 2 pears?
    6 peaches divided by 2 peaches?
    6 plums divided by 3 plums?
    8 chairs divided by 2 chairs?
    8 oranges divided by 4 oranges?

This sign ÷ stands for the words **divided by.**

4 dogs ÷ 2 = ?   6 pears ÷ 3 = ?
4 hens ÷ 2 = ?   8 cents ÷ 2 = ?
6 figs ÷ 2 = ?   8 tops ÷ 2 = ?

Divide these eight dots thus, ●●/●●/●●/●●

Into how many *equal parts* have you divided them?

If a number of things is divided into **four equal parts,** each part is **one-fourth** of the number.

How many dots in *one-fourth* of 8 dots?

How many dots in two-fourths of 8 dots?

How many dots in three-fourths of 8 dots?

How many dots in four-fourths of 8 dots?

How many dots in one-half of 8 dots?

How many dots in two-halves of 8 dots?

Fourths are often called quarters.

Find one-quarter of 4 dollars; of 8 cents.

Find two-quarters of 4 dollars; of 8 cents.

Find three-quarters of 4 dollars; of 8 cents.

Find four-quarters of 4 dollars; of 8 cents.

Find one-half of 4 dollars; of 8 cents; of 8 pigs.

What part of 8 blocks are 4 blocks? are 2 blocks?

What part of 8 cents are 2 cents? are 4 cents?

What part of 6 cups are 3 cups? are 2 cups?

Which is greater, one-half of 8 cents or one-fourth of 8 cents? one-half of 8 cents or one-quarter of 8 cents?

Which is greater, one-half of 8 cents or two-fourths of 8 cents? one-half of 8 cents or three-fourths of 8 cents?

Here is a new way of writing one-half, thus, $\frac{1}{2}$; one-third, thus, $\frac{1}{3}$; one-fourth, thus, $\frac{1}{4}$.

We write two-thirds, thus, $\frac{2}{3}$; two-fourths, thus, $\frac{2}{4}$; three-quarters, thus, $\frac{3}{4}$.

Read : $\frac{1}{2}$; $\frac{1}{3}$; $\frac{2}{3}$; $\frac{1}{4}$; $\frac{2}{4}$; $\frac{3}{4}$; $\frac{4}{4}$.

Write in figures : one-half ; one-third ; two-thirds ; one-fourth ; three-quarters.

Oral and slate exercises :

DOGS.	CATS.	PIGS.
$5 + 2 =$	$7 - 2 =$	$7 - 5 =$
$5 + 3 =$	$8 - 5 =$	$8 - 3 =$
$6 + 2 =$	$8 - 6 =$	$8 - 2 =$
$7 + 1 =$	$8 - 1 =$	$8 - 7 =$
$2 + 2 =$	$3 + 3 =$	$4 + 4 =$
$4 - 2 =$	$6 - 3 =$	$8 - 4 =$
$2 \times 2 =$	$2 \times 3 =$	$3 \times 2 =$
$4 \div 2 =$	$6 \div 3 =$	$6 \div 2 =$
$2 \times 4 =$	$4 \times 2 =$	$8 \div 2 =$
$8 \div 4 =$	$\frac{1}{2}$ of $4 =$	$\frac{1}{2}$ of $6 =$
$\frac{1}{2}$ of $8 =$	$\frac{1}{3}$ of $6 =$	$\frac{3}{4}$ of $8 =$
$\frac{1}{4}$ of $8 =$	$\frac{2}{4}$ of $8 =$	$\frac{2}{3}$ of $6 =$

HORSES.		MULES.		COLTS.	
$4 +$	$= 7.$	$5 +$	$= 8.$	$7 -$	$= 3.$
$4 +$	$= 8.$	$7 +$	$= 8.$	$8 -$	$= 4.$
$4 +$	$= 6.$	$6 +$	$= 8.$	$8 -$	$= 1.$
$4 \times$	$= 8.$	$3 +$	$= 8.$	$8 -$	$= 6.$
$2 \times$	$= 8.$	$4 +$	$= 6.$	$8 \div$	$= 4.$
$3 \times$	$= 6.$	$8 -$	$= 5.$	$6 \div$	$= 3.$
$2 \times$	$= 6.$	$8 -$	$= 7.$	$8 \div$	$= 2.$
$2 \times$	$= 4.$	$8 -$	$= 3.$	$6 \div$	$= 2.$
$5 \times$	$= 5.$	$8 -$	$= 2.$	$4 \div$	$= 2.$

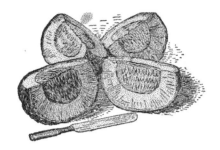

If a melon is cut into four equal parts, what is one of the parts called?

. What are three of the parts called?

How many quarters of a melon make a whole melon?

How many quarters of a melon make half of a melon?

How many quarters of a dollar make a dollar? make a half dollar?

How many quarters of a dollar make 2 dollars?

How many quarters of a dollar make one dollar and a half?

How many quarters of a dollar make one dollar and a quarter?

If a pie is cut into quarters, and Mary, Tom, and Harry each have a quarter, how many quarters will be left for Alice?

If half of a pie is cut into two equal parts, what part of the *whole pie* is each piece?

What part of the whole pie are the two pieces together?

How many fourths make one half?

How many fourths make one whole?

GILL.   PINT.   QUART.   GALLON.

Which one of these measures is the smallest?
How many gills will the pint measure hold? *
Then four gills make one pint.
At 1 cent a gill, what will a pint of milk cost?
At 2 cents a gill, what will a pint of syrup cost?
How many gills in a half pint of water?

How many pints will the quart measure hold? *
Then two pints make one quart.
At 4 cents a pint, what will a quart of milk cost?
At 3 cents a pint, what will a quart of oil cost?
At 6 cents a quart, what will a pint of berries cost?
What part of a quart is 1 pint?
How many gills make 1 quart?

How many quarts will the gallon measure hold? *
Then four quarts make one gallon.
How many quart cans are needed for a gallon
of milk?   How many two-quart cans?
At 2 cents a quart, what will a gallon of skim-
milk cost?   What will a half-gallon cost?
What part of a gallon is one quart?
What part of a gallon are 2 quarts? are 3 quarts?
At 8 cents a gallon, what will a quart of skim-
milk cost?   What will a pint cost?

Note.* Let the pupil discover by trial the answer to this question.

THE NUMBER NINE.

Eight dots and *one* dot make **nine** dots.

Here are *nine* dots, ⣿

We write the **figure** 9 for **nine**.

Copy these pictures, and write under each group the figure for the number in the group:

Copy these cards, and add dots enough to make 9 dots in each card, and write 9 under each card:

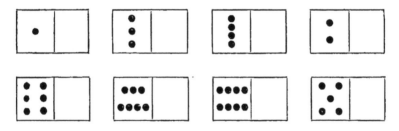

How many dots are 3 and 6? 5 and 4? 7 and 2? 1 and 8? 3 and 3? 2 and 7? 4 and 3? 4 and 5? 4 and 4? 6 and 3? 8 and 1? 5 and 3?

How many dots must you put with 5 to make 9?

How many dots must you put with 2 to make 9?

How many dots must you put with 3 to make 9?

How many dots must you put with 4 to make 9?

How many dots must you put with 6 to make 9?

How many dots must you put with 8 to make 9?

How many dots must you put with 7 to make 9?

How many more dots are 9 than 7? 9 than 6? 9 than 3? 9 than 4? 9 than 5? 9 than 2?

Here are 9 blocks,

If you take away 3 blocks, how many will be left?
If you take away 6 blocks, how many will be left?
If you take away 5 blocks, how many will be left?
If you take away 4 blocks, how many will be left?
If you take away 7 blocks, how many will be left?
If you take away 2 blocks, how many will be left?
If you take away 1 block, how many will be left?
If you take away 8 blocks, how many will be left?
How many are :

8 minus 2 ?	9 minus 3 ?	7 minus 4 ?	9 minus 4 ?
8 minus 6 ?	9 minus 6 ?	7 minus 6 ?	9 minus 5 ?
8 minus 5 ?	9 minus 7 ?	7 minus 3 ?	8 minus 3 ?
8 minus 4 ?	9 minus 8 ?	7 minus 2 ?	9 minus 1 ?
8 minus 7 ?	9 minus 2 ?	7 minus 5 ?	6 minus 4 ?

Emma may take 3 blocks at a time for 3 times.

How many blocks has Emma?

How many blocks, then, are 3 times 3 blocks?

How many peaches are 3 times 3 peaches?

How many roses are 3 times 3 roses?

How many lambs are 3 times 3 lambs?

How many are 3 times 3?

Here are 9 pears,

How many times can you take 3 pears from the 9?

How many groups of 3 pears each can you make from the 9?

9 pears divided by 3 pears gives how many times?

9 pears divided by 3 gives how many pears?

How many pears are $\frac{1}{3}$ of 9 pears?

Here are 9 dots, ● ● ● ● ● ● ● ● ●

Put your pencil between the second and third dots. How many dots are on the left of the pencil? How many on the right of the pencil?

Put your pencil between the fourth and fifth dots. How many dots are on the left of the pencil? How many on the right of the pencil?

Put your pencil between the fifth and sixth dots. How many dots on the left? How many dots on the right?

If a boy goes up 8 steps 2 steps at a time, how many steps will he touch?

John had 9 flags, some of them red and the rest blue. If 4 of them were red, how many were blue?

Alice had 9 cents, and spent 3 cents. How many had she left?

George sells a newspaper for 2 cents, and receives a five-cent piece in payment. How many cents must he give back?

A hen had 9 chickens, but a hawk caught 2. How many chickens were left?

Miriam has 8 cents, and Hattie 3 less than Miriam. How many cents has Hattie?

Harry has 3 tops, and Tom has 6 more than Harry. How many tops has Tom?

I wanted 8 stamps for my letters, and had only 3. How many more must I buy?

Tom had 6 apples, and gave away ⅓ of them. How many had he left?

Florence had 8 apples on plates, 2 on a plate. How many plates were there ?   How many twos in eight ?

Annie had 8 pears on plates, 4 on a plate.   How many plates were there ?   How many fours in eight ?

Hattie had 9 peaches, 3 on a plate.   How many plates were there ?   How many threes in 9 ?

$$8 \div 2 = \qquad 8 \div 4 = \qquad 9 \div 3 =$$
$$\tfrac{1}{2} \text{ of } 8 = \qquad \tfrac{1}{4} \text{ of } 8 = \qquad \tfrac{1}{3} \text{ of } 9 =$$

Mary had 3 rows of buttons, 3 in a row.   How many buttons had she ?

If one orange costs 3 cents, what will 2 oranges cost ?   What will 3 oranges cost ?

If a quart of milk costs 6 cents, what will a pint cost ?   What will 3 pints cost ?

If a pint of vinegar costs 4 cents, what will a gill cost ?   What will a quart cost ?

If a pint of water will fill 4 gill cups, how many gill cups will a quart of water fill ?

If a quart of milk will fill 2 pint cups, how many pint cups will a gallon of milk fill ?

How many quart measures will a one-gallon can of milk fill ? will a two-gallon can fill ?

A cook had 9 eggs, and used $\tfrac{1}{3}$ of them for a pudding.   How many eggs were left ?

Harry had 8 oranges.   He gave one-quarter of them to his sister Mary, one-quarter of them to his sister Alice, and one-quarter of them to his sister Ellen.   How many did he keep for himself ?

Oral and slate exercises :

FIGS.	BELLS.	APPLES.
$2 + 7 =$	$9 - 2 =$	$9 - 7 =$
$4 + 5 =$	$9 - 4 =$	$9 - 5 =$
$3 + 5 =$	$8 - 3 =$	$8 - 5 =$
$3 + 6 =$	$9 - 3 =$	$9 - 6 =$
$5 + 2 =$	$7 - 5 =$	$7 - 2 =$
$4 + 3 =$	$7 - 3 =$	$7 - 4 =$
$1 + 8 =$	$9 - 6 =$	$9 - 8 =$
$4 + 2 =$	$6 - 4 =$	$6 - 2 =$
$4 + 4 =$	$6 - 3 =$	$8 - 4 =$
$3 \times 3 =$	$\frac{1}{2}$ of $4 =$	$\frac{1}{2}$ of $6 =$
$\frac{1}{2}$ of $2 =$	$\frac{1}{3}$ of $3 =$	$\frac{1}{3}$ of $6 =$
$\frac{1}{3}$ of $9 =$	$\frac{1}{4}$ of $8 =$	$\frac{1}{2}$ of $8 =$

SLEDS.	ORANGES.	LAMBS.
$4 = 1 +$	$6 = 4 +$	$8 = 3 +$
$4 = 2 +$	$7 = 1 +$	$8 = 4 +$
$4 = 3 +$	$7 = 4 +$	$8 = 5 +$
$5 = 1 +$	$7 = 5 +$	$9 = 1 +$
$5 = 3 +$	$7 = 3 +$	$9 = 8 +$
$5 = 4 +$	$7 = 6 +$	$9 = 2 +$
$5 = 2 +$	$7 = 2 +$	$9 = 7 +$
$6 = 3 +$	$8 = 1 +$	$9 = 3 +$
$6 = 1 +$	$8 = 6 +$	$9 = 6 +$
$6 = 2 +$	$8 = 7 +$	$9 = 4 +$
$6 = 5 +$	$8 = 2 +$	$9 = 5 +$

NOTE. Besides copying and completing these and similar exercises, the oral drill must be kept up until *every one* of the class can give the answers promptly.

THE FIGURE ZERO.

The figure 0 is called **zero, naught,** or **cipher.**

The figure 0 means **none to count.**

Four roses grew on a bush; 2 were picked; and then 2 more. Write the figure for the number left.

Write the figure for the number of blocks left when you take away 5 blocks from 5 blocks.

Write the figure for the number of oranges left if you had 4 oranges and gave them all away.

Review of figures:

You have now had *all* the figures used for writing numbers, and have learned the meaning of each separate figure. Thus:

> The figure 1 is written for **One.**
> The figure 2 is written for **Two.**
> The figure 3 is written for **Three.**
> The figure 4 is written for **Four.**
> The figure 5 is written for **Five.**
> The figure 6 is written for **Six.**
> The figure 7 is written for **Seven.**
> The figure 8 is written for **Eight.**
> The figure 9 is written for **Nine.**
> The figure O is written for **None.**

Draw a square, and write 0 under it: then another square, and write 1 under it; and so on to 9.

Put in each square the number of dots indicated by the figure written under it

THE NUMBER TEN.

Nine dots and *one* dot make **ten** dots.

Here are *ten* dots,

We write the **figures 10** for **ten**.

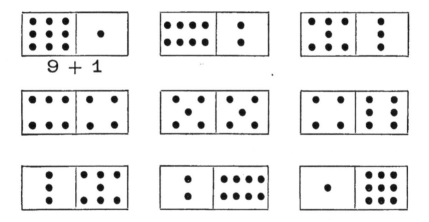

Ten ones make **1 ten.**

Draw these number pictures of ten, and write under each division the figure for the number of dots in the division :

9 + 1

Look at these number cards, and answer the following questions :

How many must you add to 9 to make 10 ? to 8 to make 10 ? to 7 to make 10 ? to 6 to make 10 ? to 5 to make 10 ? to 4 to make 10 ? to 3 to make 10 ? to 2 to make 10 ? to 1 to make 10 ?

How many more are 10 than 2 ? than 4 ? than 6 ? than 8 ? than 3 ? than 5 ? than 7 ? than 9 ?

John found 3 eggs on Thursday, and 4 on Friday. How many eggs did he find?

There are 7 days in a week. When 2 are gone, how many are left?

There were 7 red roses on a rose bush, and 5 white roses on another bush. How many more red roses were there than white roses?

Susan had 3 apples, and James had 5 apples How many apples had Susan and James together?

Alice had 8 dolls, and gave away 3 of them. How many had she left?

If a window has 4 panes of glass, how many panes of glass in 2 windows?

How many feet have 2 dogs? 4 hens?

If you take 2 apples 4 times from a dish that has 8 apples in it, how many apples will be left?

There were 8 rooms in a house, half of them in the first story, and half in the second story. How many were there in each story?

John had 8 peaches, and gave away a quarter of them. How many peaches did he give away?

Frank bought 9 marbles, and gave away 4 of them. How many had he left?

There are 9 apples in a dish. How many boys can have 3 apples apiece?

At 3 cents apiece, how many oranges can you buy for 6 cents? for 9 cents?

At 2 cents apiece, how many pears can you buy for 9 cents, and how many cents will be left?

Part II.

LESSON 1.

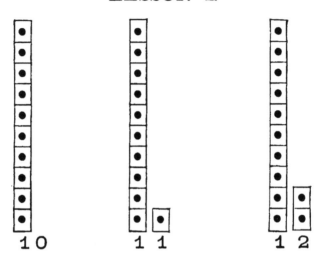

Look at the number picture on the right. What do you see over the 2? **2 ones.** Over the 1? **1 ten.** Then 12 means *one ten* and *two ones.*

Look at the middle number. What do you see over the 1 at the right? What do you see over the 1 at the left? Then 11 means *one ten* and *one.*

Look at the number picture on the left. What do you see over the 0? What do you see over the 1? Then 10 means *one ten* and *no ones.*

NOTE. The Teacher should proceed in Part II. as in Part I.; showing objects, drawing number pictures on the board, *and reading all the clothed exercises for the pupils.* Pupils should have the books simply to copy and solve the numerical exercises.

John may go to the counting-board. How many holes are there in the top row? Put one nail in one of the holes of the top row.

How many holes are left in the row?

Then how many must we add to 1 to make 10?

Put in one more nail. How many nails are there now? How many holes are left in the row?

Then how many must we add to 2 to make 10?

Put in one more nail. How many nails are there now? How many holes are left in the row?

How many must we add to 3 to make 10?

Put in one more nail. How many nails are there now? How many holes are left in the row?

How many must we add to 4 to make 10?

Put in one more nail. How many nails in the row? How many holes are left in the row?

How many must we add to 5 to make 10?

Put in one more nail. How many nails in the row? How many holes are left in the row?

How many must we add to 6 to make 10?

Put in one more nail. How many nails in the row? How many holes are left in the row?

How many must we add to 7 to make 10?

Put in one more nail. How many nails are there now? How many holes are left in the row?

How many must we add to 8 to make 10?

Put in one more nail. How many nails now in the row? How many holes are left in the row?

How many must we add to 9 to make 10?

Here are ten rings, ⦶ ◯ ◯ ◯ ◯ ◯ ◯ ◯ ◯ ◯
I will put the end of the pointer between the second and third rings.

How many rings on the left of the pointer?

How many rings on the right of the pointer?

How many are 2 and 8 ?

How many are 10 less 2? 10 less 8 ?

I will put the end of the pointer between the third and fourth rings.

How many rings on the left of the pointer?

How many rings on the right of the pointer?

How many are 3 and 7 ?

How many are 10 less 3? 10 less 7 ?

I will put the end of the pointer between the fourth and fifth rings.

How many rings on the left of the pointer?

How many rings on the right of the pointer?

How many are 4 and 6 ?

How many are 10 less 4? 10 less 6 ?

I will put the end of the pointer between the fifth and sixth rings.

How many rings on the left of the pointer?

How many rings on the right of the pointer?

How many are 5 and 5 ?

How many are 10 less 5 ?

How many are 10 less 1? 10 less 9 ?

NOTE. Practise this exercise, putting the pointer in different posi‑tions, until the pupils can readily name any two parts of 10, and the part left when one part is taken from 10.

In each of the number pictures below, the bundle is a bundle of ten.

Write the figures for the number in each case.

How many figures do you write for each number?
What does the figure on the left show?
What does the figure on the right show?
What is the number 11 called? Eleven.
What is the number 12 called? Twelve.

NOTE. It is absolutely necessary for the Teacher to show bundles of ten things (pencils, sticks, etc.) kept distinct by rubber bands, in order to show the compositions of numbers containing tens and ones; and to show also that the counting of **units of tens** is exactly the same as the counting of single units.

Oral and slate exercises:

ROBINS.	ROBINS.	ROBINS.
$8 + ? = 10.$	$5 = 1 + ?$	$7 = 5 + ?$
$6 + ? = 10.$	$5 = 2 + ?$	$7 = 3 + ?$
$5 + ? = 10.$	$4 = 2 + ?$	$8 = 2 + ?$
$1 + ? = 10.$	$4 = 1 + ?$	$8 = 3 + ?$
$3 + ? = 10.$	$6 = 1 + ?$	$8 = 4 + ?$
$7 + ? = 10.$	$6 = 3 + ?$	$9 = 3 + ?$
$2 + ? = 10.$	$6 = 4 + ?$	$9 = 5 + ?$
$4 + ? = 10.$	$7 = 6 + ?$	$9 = 2 + ?$
$9 + ? = 10.$	$7 = 4 + ?$	$9 = 8 + ?$

NOTE. Continue these oral and slate exercises until every pupil can separate 10 into any two parts, and see at a glance the number to be added to any part to make 10; and also see the part required when a number less than 10 and one of its parts is given.

There were 5 birds in a tree, and 5 more flew in the tree. How many birds were in the tree then?

A teamster has 5 teams of 2 horses each. How many horses has he?

Harry brought in some wood twice. The first time he brought in 4 sticks, and the next time 5 sticks. How many sticks did he bring in?

There are 4 plates on each side of a table, and one plate at each end. How many plates in all?

If a table is 3 feet long and 2 feet wide, how many feet long are the 2 sides and 2 ends together?

A farmer brought 10 bushels of potatoes to put into his cellar. After he had put in 6 bushels, how many more bushels remained to be put in?

Daisy has 10 chickens. Five are white, and the rest brown. How many are brown?

A room is 10 feet high, and the top of the door is 7 feet from the floor. How many feet from the top of the door is the ceiling?

There were 10 saucers and only 8 cups. How many saucers were without cups?

I have 10 letters to mail, and only 1 stamp. How many stamps must I buy?

If a boy has 10 apples, and eats 2 apples a day, how many days will they last?

If a boy has 10 cents, and spends half of them, how many will he have left?

NOTE. These and similar questions can be made more intelligible and interesting by illustrating them with suitable number pictures.

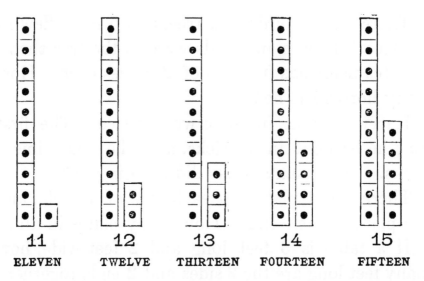

How many dots are 10 dots and 1 dot? 10 dots and 2 dots? 10 dots and 3 dots? 10 dots and 4 dots? 10 dots and 5 dots?

How many sheep are 10 sheep and 1 sheep? 10 sheep and 2 sheep? 10 sheep and 3 sheep? 10 sheep and 4 sheep? 10 sheep and 5 sheep?

If you have 10 oranges, how many more must you buy to have 13? to have 14?

How many blocks must you add to 10 blocks to have 15? to have 12? to have 11?

How many twos are there in 8? in 10?

Oral and slate exercises:

TOPS.	BALLS.	CHICKENS.
$10 + 1 = ?$	$11 - 1 = ?$	$11 - 10 = ?$
$10 + 3 = ?$	$13 - 3 = ?$	$13 - 10 = ?$
$10 + 5 = ?$	$15 - 5 = ?$	$15 - 10 = ?$
$10 + 2 = ?$	$12 - 2 = ?$	$12 - 10 = ?$
$10 + 4 = ?$	$14 - 4 = ?$	$14 - 10 = ?$

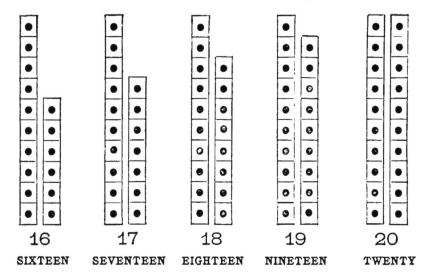

16	17	18	19	20
SIXTEEN	SEVENTEEN	EIGHTEEN	NINETEEN	TWENTY

How many dots are 10 dots and 6 dots? 10 dots and 7 dots? 10 dots and 8 dots? 10 dots and 9 dots? 10 dots and 10 dots?

If you have 10 cards, how many more must you have to make 16 cards? to make 17 cards?

How many marbles must you put with 10 marbles to make 19 marbles? to make 18 marbles?

How many cents have you if you have 10 cents, 5 cents, and 1 cent?

How many cents have you if you have ten cents, five cents, and two cents?

Oral and slate exercises:

CROWS.	CROWS.	CROWS.
$10 + 7 = ?$	$17 - 7 = ?$	$17 - 10 = ?$
$10 + 9 = ?$	$19 - 9 = ?$	$19 - 10 = ?$
$10 + 6 = ?$	$16 - 6 = ?$	$16 - 10 = ?$
$10 + 8 = ?$	$18 - 8 = ?$	$18 - 10 = ?$
$10 + 10 = ?$	$20 - 10 = ?$	$15 - 10 = ?$

Write under the number pictures below the figures for the number, and the name of the number.

In which place do we write the **ones**? the tens?

Note. Pupils should be made familiar with the dime and all coins of smaller value; and with the ten-cent postage stamp, and all stamps of smaller value.

Annie has 2 five-cent pieces and a one-cent piece. How much money has Annie? $2 \times 5 + 1 = ?$

What two pieces of money together make 11 cents? $10 + 1 = ?$

What two pieces of money together make 12 cents? $10 + 2 = ?$

What two pieces of money together make 15 cents? $10 + 5 = ?$

Alice has 2 five-cent pieces and a two-cent piece. How much money has Alice? $2 \times 5 + 2 = ?$

Harry has 3 five-cent pieces. How much money has Harry? $5 + 5 + 5 = ?$

What five pieces of money together make 14 cents? $5 + 5 + 2 + 1 + 1 = ?$

What three pieces of money together make 13 cents? $10 + 2 + 1 = ?$

What four pieces of money together make 13 cents? $5 + 5 + 2 + 1 = ?$

What four pieces of money together make 14 cents? $5 + 5 + 2 + 2 = ?$

Write under the number pictures below the figures for the number, and the name of the number.

In which place do we write the **ones**? the **tens**?

How many tens and how many ones are there in 16? in 17? in 18? in 19?

How many ones must we add to 9 *ones* to make 1 *ten?* to 7 *ones* to make 1 *ten?* to 6 *ones* to make 1 *ten?* to 8 *ones* to make 1 *ten?*

How many tens and how many ones in 20?

What does the figure 0 mean in the number 20?

How many twos in 10? XX XX XX XX XX

How many fives in 10? XXXXX XXXXX

How many more twos in 16 than in 10? How many twos in 16?

How many more twos in 18 than in 10? How many twos in 18? How many twos in 12?

How many more ones in 19 than in 17?

How many more ones in 19 than in 16?

How many more ones in 19 than in 10?

How many more ones in 18 than in 10?

How many more ones in 16 than in 10?

How many more ones in 17 than in 10?

How many more ones in 18 than in 16?

How many more ones in 15 than in 10?

How many more ones in 15 than in 12?

Oral and slate exercises :

SHEEP.	LAMBS.	MEN.
$12 + 2 = ?$	$11 + 2 = ?$	$13 + 3 = ?$
$11 + 4 = ?$	$15 + 2 = ?$	$14 + 3 = ?$
$14 + 5 = ?$	$13 + 4 = ?$	$12 + 6 = ?$
$16 + 3 = ?$	$14 + 2 = ?$	$17 + 1 = ?$
$12 + 4 = ?$	$12 + 5 = ?$	$11 + 8 = ?$
$13 + 6 = ?$	$11 + 7 = ?$	$15 + 3 = ?$
$15 + 4 = ?$	$17 + 2 = ?$	$13 + 5 = ?$
$18 + 1 = ?$	$12 + 3 = ?$	$11 + 3 = ?$
$12 + 7 = ?$	$11 + 6 = ?$	$14 + 4 = ?$

EGGS.	HENS.	DUCKS.
$17 - 1 = ?$	$15 - 3 = ?$	$14 - 2 = ?$
$13 - 2 = ?$	$19 - 4 = ?$	$18 - 6 = ?$
$19 - 5 = ?$	$19 - 7 = ?$	$16 - 5 = ?$
$16 - 2 = ?$	$14 - 2 = ?$	$19 - 4 = ?$
$19 - 6 = ?$	$17 - 3 = ?$	$18 - 4 = ?$
$14 - 3 = ?$	$19 - 2 = ?$	$16 - 4 = ?$
$15 - 2 = ?$	$17 - 4 = ?$	$19 - 8 = ?$
$17 - 5 = ?$	$18 - 5 = ?$	$17 - 2 = ?$
$16 - 3 = ?$	$19 - 3 = ?$	$18 - 1 = ?$

NOTE. The above exercises, and similar exercises, should be worked aloud by *each one* of the class in turn ; and on blocks of paper or slates. Thus, the first example should be worked at first, as follows : 12 sheep and 2 sheep are 14 sheep.

If a child makes a mistake, let the child himself correct it by the counting-board or by dots on the blackboard. Care should be taken to have him clearly see that these operations are confined to the *ones*. Thus, in adding 2 to 12, let him fill the top row of holes in the counting-board with nails, and 2 holes more in the next row for the 12, then put two more nails in the row with the 2 nails already there. He will then see that $12 + 2 = 10 + 4 = 14$.

How many cents are 12 cents and 5 cents ?

How many days are 1 week and 3 days ?

How many inches are 11 inches and 7 inches ?

How many boys are 13 boys and 6 boys ?

How many pinks are 15 pinks and 3 pinks ?

One rose-bush has 17 roses, and another only 2. How many have both bushes together ? How many more has one bush than the other ?

A farmer has 16 cows in the barn, and 3 in the stable. How many cows has he in all ? How many more in the barn than in the stable ?

A man has 14 work horses and 2 driving horses. How many horses has he ? How many more work horses than driving horses ?

James found 15 eggs in one nest, and 5 in another. How many eggs did he find in both nests ?

The number 12 is sometimes called a dozen.

When we say a *dozen eggs,* we mean *twelve* eggs.

Frank started with a dozen eggs from the barn, but dropped and broke two before he reached the house. How many did he carry into the house ?

John has a dozen chickens of one kind, and 6 of another kind. How many has he of both kinds ?

Harry had a dozen oranges, but he gave away ten. How many had he left ?

A watch dealer had 3 dozen gold watches the week before Christmas ; the day after Christmas he had 1 dozen left. How many dozen had he sold ? How many watches had he left ?

Oral and slate exercises :

CROWS.	ROBINS.
$9 + 3 = 10 + 2 = 12.$	$7 + 7 = 10 + 4 = 14.$
$9 + 8 = 10 + ? = ?$	$7 + 4 = 10 + ? = ?$
$9 + 4 = 10 + ? = ?$	$7 + 8 = 10 + ? = ?$
$9 + 6 = 10 + ? = ?$	$6 + 6 = 10 + ? = ?$
$9 + 5 = 10 + ? = ?$	$6 + 5 = 10 + ? = ?$
$9 + 7 = 10 + ? = ?$	$6 + 7 = 10 + ? = ?$
$9 + 2 = 10 + ? = ?$	$6 + 9 = 10 + ? = ?$
$9 + 9 = 10 + ? = ?$	$6 + 8 = 10 + ? = ?$
$8 + 3 = 10 + ? = ?$	$5 + 9 = 10 + ? = ?$
$8 + 5 = 10 + ? = ?$	$5 + 7 = 10 + ? = ?$
$8 + 7 = 10 + ? = ?$	$5 + 8 = 10 + ? = ?$
$8 + 6 = 10 + ? = ?$	$5 + 6 = 10 + ? = ?$
$8 + 4 = 10 + ? = ?$	$4 + 8 = 10 + ? = ?$
$8 + 8 = 10 + ? = ?$	$4 + 7 = 10 + ? = ?$
$8 + 9 = 10 + ? = ?$	$4 + 9 = 10 + ? = ?$
$7 + 5 = 10 + ? = ?$	$3 + 8 = 10 + ? = ?$

NOTE. When the sum of the ones is more than ten, we proceed as follows : Suppose we have to add 7 to 8. Call upon one of the children to put 8 nails in the top row of the counting-board, and 7 in the second row, and then ask, How many nails are there in the top row ? How many holes are left ? How many nails must we put in the top row to make ten ? Let him take 2 nails from the 7 in the second row and put in the holes left in the top row. How many nails now in the top row ? How many in the second row ? Then 8 and 7 are 10 and 5, or 15.

Continue this practice, a few minutes at a time, until the children can dispense with the counting-board ; then continue it with the intermediate step until they can dispense with that step, and name instantly the sum of any two numbers that are each less than ten.

This method may seem tedious, but it is the only method that gives *complete mastery of addition.*

Oral and slate exercises :

SPARROWS.	KINGBIRDS.
$8 + 6 = 10 + ? = ?$	$4 + 7 = 10 + ? = ?$
$7 + 4 = 10 + ? = ?$	$8 + 4 = 10 + ? = ?$
$5 + 8 = 10 + ? = ?$	$7 + 8 = 10 + ? = ?$
$8 + 7 = 10 + ? = ?$	$2 + 9 = 10 + ? = ?$
$9 + 3 = 10 + ? = ?$	$9 + 4 = 10 + ? = ?$
$8 + 5 = 10 + ? = ?$	$9 + 9 = 10 + ? = ?$
$6 + 5 = 10 + ? = ?$	$8 + 8 = 10 + ? = ?$
$5 + 7 = 10 + ? = ?$	$7 + 7 = 10 + ? = ?$
$4 + 9 = 10 + ? = ?$	$6 + 6 = 10 + ? = ?$
$5 + 8 = 10 + ? = ?$	$8 + 9 = 10 + ? = ?$
$7 + 6 = 10 + ? = ?$	$3 + 9 = 10 + ? = ?$
$7 + 9 = 10 + ? = ?$	$2 + 9 = 10 + ? = ?$

DUCKS.	TURKEYS.	CHICKENS.
$9 + 9 = ?$	$8 + 9 = ?$	$6 + 8 = ?$
$9 + 7 = ?$	$8 + 8 = ?$	$6 + 4 = ?$
$9 + 4 = ?$	$7 + 3 = ?$	$5 + 5 = ?$
$9 + 2 = ?$	$7 + 7 = ?$	$5 + 8 = ?$
$9 + 6 = ?$	$7 + 5 = ?$	$5 + 6 = ?$
$9 + 3 = ?$	$7 + 8 = ?$	$5 + 7 = ?$
$9 + 5 = ?$	$7 + 6 = ?$	$5 + 9 = ?$
$9 + 8 = ?$	$7 + 4 = ?$	$4 + 7 = ?$
$8 + 3 = ?$	$7 + 9 = ?$	$4 + 9 = ?$
$8 + 5 = ?$	$6 + 6 = ?$	$4 + 8 = ?$
$8 + 4 = ?$	$6 + 9 = ?$	$3 + 8 = ?$
$8 + 7 = ?$	$6 + 7 = ?$	$3 + 9 = ?$
$8 + 6 = ?$	$6 + 5 = ?$	$2 + 9 = ?$

How many days are 1 week and 4 days? 1 week and 5 days? 1 week and 6 days? 2 weeks?

John has 9 cents, and Mary 4 cents. How many have both? How many are 9 and 4? 4 and 9?

If one lamp is worth 6 dollars, and another 5 dollars, how much are both worth?

If there are 8 boys in one class, and 5 in another, how many are there in both classes?

If there are 6 boys in one class, and 7 in another, how many are there in both classes?

A farmer sold 6 sheep to one man, and 8 to another. How many sheep did he sell?

A farmer has 9 cows in one pasture, and 5 in another. How many cows has he in the two pastures? How many are 9 and 5? 5 and 9?

Tom has two hens, one white, and the other black. The white hen has 9 chickens, and the black hen has 8 chickens. How many chickens have both hens? How many are 9 and 8? 8 and 9?

James saw 9 crows on the ground, and 7 more flying about. How many crows did he see?

There are 8 blocks in one pile, and 8 in another pile. How many blocks are there in both piles?

There were 9 chickens roosting on one pole, and 6 on another pole. How many chickens were roosting on both poles? How many are 9 and 6?

If Harry paid 8 cents for his block of paper, and Ernest paid 7 cents for his, how many cents did the two blocks cost?

Oral and slate exercises :

CHAIRS.	BOXES.

$11 - 2 = 10 - 1 = 9.$ $13 - 7 = 10 - 4 = 6.$

$11 - 3 = 10 - ? = ?$ $13 - 8 = 10 - ? = ?$

$11 - 4 = 10 - ? = ?$ $13 - 5 = 10 - ? = ?$

$11 - 5 = 10 - ? = ?$ $14 - 5 = 10 - ? = ?$

$11 - 6 = 10 - ? = ?$ $14 - 6 = 10 - ? = ?$

$11 - 7 = 10 - ? = ?$ $14 - 7 = 10 - ? = ?$

$11 - 8 = 10 - ? = ?$ $14 - 8 = 10 - ? = ?$

$11 - 9 = 10 - ? = ?$ $14 - 9 = 10 - ? = ?$

$12 - 3 = 10 - ? = ?$ $15 - 6 = 10 - ? = ?$

$12 - 4 = 10 - ? = ?$ $15 - 7 = 10 - ? = ?$

$12 - 5 = 10 - ? = ?$ $15 - 8 = 10 - ? = ?$

$12 - 6 = 10 - ? = ?$ $15 - 9 = 10 - ? = ?$

$12 - 7 = 10 - ? = ?$ $16 - 7 = 10 - ? = ?$

$12 - 8 = 10 - ? = ?$ $16 - 8 = 10 - ? = ?$

$12 - 9 = 10 - ? = ?$ $16 - 9 = 10 - ? = ?$

$13 - 4 = 10 - ? = ?$ $17 - 8 = 10 - ? = ?$

$13 - 5 = 10 - ? = ?$ $17 - 9 = 10 - ? = ?$

$13 - 6 = 10 - ? = ?$ $18 - 9 = 10 - ? = ?$

NOTE. In Subtraction, the pupils may use the knowledge acquired in Addition. Thus, if 8 is to be subtracted from 15, the answer sought is obtained by discovering the number that must be added to 8 to make 15. But it is better to keep Subtraction distinct from Addition, and at this stage to take two steps, just as we did in Addition.

Suppose we are required to take 8 from 15. Let one of the children put 10 nails in the top row of holes in the counting-board, and 5 in the next row below. We now ask the following questions: How many nails must we take away to leave 10? How many more than 5 are we required to take away? And 3 nails from 10 nails leave? Then $15 - 8 = 10 - 3 = 7.$

Oral and slate exercises :

BUTTONS.	NEEDLES.	PINS.
$12 - 3 =$	$14 - 8 =$	$11 - 6 =$
$13 - 6 =$	$12 - 6 =$	$15 - 7 =$
$11 - 5 =$	$11 - 3 =$	$13 - 4 =$
$15 - 9 =$	$16 - 8 =$	$13 - 7 =$
$16 - 7 =$	$13 - 9 =$	$12 - 5 =$
$13 - 8 =$	$15 - 8 =$	$11 - 8 =$
$11 - 7 =$	$17 - 9 =$	$14 - 7 =$
$12 - 9 =$	$11 - 2 =$	$12 - 8 =$
$15 - 6 =$	$12 - 7 =$	$16 - 9 =$
$14 - 6 =$	$14 - 5 =$	$18 - 9 =$
$11 - 4 =$	$12 - 4 =$	$13 - 5 =$
$11 - 9 =$	$16 - 8 =$	$14 - 9 =$

If you pay 17 dollars for a table, and 8 dollars for a chair, how many dollars more do you pay for the table than for the chair?

John has 16 marbles, and James has 9. How many more has John than James?

Take 1 week from 14 days. How many days are left? How many weeks are left?

I have 17 miles to walk. After I have walked 9 miles, how many more have I to walk?

A milkman has 16 cows. If he sells 7, how many will be left?

A farmer had 16 turkeys, but a fox carried off 8 of them. How many were left for the farmer?

Alice has 15 chickens.   If 6 are black, and the rest are white, how many are white?

If Ernest had 9 marbles more, he would have 15. How many marbles has he?

The first train in the morning had 7 cars, and the second train had 15 cars.   How many more cars did the second train have than the first train?

Mary picked 15 quarts of blueberries, and George picked 8 quarts.   How many more quarts did Mary pick than George?

George caught 14 trout, and his brother caught 8 trout.   How many more did George catch than his brother?

Henry had 14 cents, but spent 6 cents for lemons.   How many cents had he left?

Miriam is 14 years old.   How old was she 7 years ago? 9 years ago?

Lucy's father and mother together gave her 14 cents.   Her father gave her 9 cents.   How many cents did her mother give her?

There were 14 rolls on the table before breakfast, and only 5 after breakfast.   How many rolls were eaten at breakfast?

Frank had 13 cents.   He had one five-cent piece, and the rest one-cent pieces.   How many one-cent pieces did he have?

Mary's mother had 13 eggs.   She used 4 for a pudding.   How many were left?

How many are 14 minus 6? 15 minus 8?

I sent by mail two books, and paid 13 cents postage. The postage for one was 8 cents. How much was the postage for the other?

A farmer had 13 lambs. How many had he left if he sold 6? if he sold 7?

Tom had 13 oranges, but he gave away 9. How many had he left?

Edna's class numbers 12. If 5 are boys, how many are girls?

Harry had 12 papers to sell. After he had sold 9, how many had he to sell?

Lucy had 12 plums, and Alice had 4. How many more had Lucy than Alice?

In two pods there were 12 peas. If there were 6 in one pod, how many were there in the other?

Erwin found a nest of 12 eggs. If he carried 3 of the eggs into the house, how many were left?

Fred had 12 cents. How many had he left if he spent 8 cents? if he spent 7 cents?

Jane bought 11 yards of ribbon, and used 6 yards. How many yards had she left?

Lucy is 11 years old, and Mary 7. How many years older is Lucy than Mary?

Frank bought 3 oranges for 9 cents, and sold them for 11 cents. How many cents did he gain?

Grace had 11 cents, and paid 5 cents for car-fare. How many cents had she left?

There were 11 saucers on the table, but 3 had no cups. How many had cups?

TWELVE. 12.

(a)

(b)

Look at the number picture marked (a).

How many dots are there in each row?

How many rows are there?

How many dots in the three rows?

Then how many are 3 times 4 dots?

A line of dots running up and down the page is called a **column**.

How many dots in each column?

How many columns are there?

How many dots in the four columns?

Then how many are 4 times 3 dots?

How many 3's in 12?   How many 4's in 12?

Look at the number picture marked (b).

How many dots are there in each row?

How many rows are there?

How many dots in the two rows?

Then how many are 2 times 6 dots?

How many dots are there in each column?

How many columns are there?

How many dots in the six columns?

Then how many are 6 times 2 dots?

How many 2's in 12?   How many 6's in 12?

Find $\frac{1}{2}$ of 12; $\frac{1}{3}$ of 12; $\frac{1}{4}$ of 12; $\frac{1}{6}$ of 12.

$12 \div 2 = ?$	$12 \div 3 = ?$	$12 \div 4 = ?$	$12 \div 6 = ?$
$2 \times 3 = ?$	$2 \times 4 = ?$	$2 \times 5 = ?$	$2 \times 6 = ?$
$3 \times 3 = ?$	$3 \times 4 = ?$	$4 \times 3 = ?$	$6 \times 2 = ?$

THE FOOT-RULE AND THE YARD-STICK.

Measure the yard-stick with the foot-rule. How many feet long is the yard-stick?

A carpet is a yard wide. How many feet wide is the carpet?

How many yards in 3 feet? in 6 feet? in 9 feet? in 12 feet?

How many feet in 2 yards? in 3 yards? in 4 yards? in ⅓ of a yard?

If the distance between two windows is 3 yards, how many feet is the distance?

Your foot-rule is marked off into 12 divisions. What is each division called? How many inches, then, make a foot?

How many inches in ½ a foot? ⅓ of a foot? ¼ of a foot? ⅔ of a foot? ¾ of a foot?

What part of a foot are 6 inches? 4 inches?

How many more inches are 10 inches than 6 inches? than 5 inches? than 3 inches? than 7 inches? than 2 inches?

Remember: **12 inches make 1 foot.**
            3 feet   make 1 yard.

If there are 8 yards of wall-paper in a roll, how many yards are there in ½ of a roll?

If it takes 2 yards of ribbon to trim a hat, how many yards will it take to trim 6 hats?

Edna's mother had 8 yards of velvet. She used ¼ of her velvet. How many yards were left?

Measure with the foot rule:

1. The length of a page of your reader.
2. The length of the top of your desk.
   The length of a pane of glass in the window.
   . The width of a pane of glass in the window.
   The length of your slate.
   The width of your slate.
   The length of the face of the blackboard.
   . The width of the face of the blackboard.
9. The length of a page of your copybook.

Measure with the yard stick:

10. The width of the floor of this room.
11. The length of the floor of this room.

Draw on the board a line 12 inches long, as nearly as you can without measuring. Measure this line, and tell me how long it really is.

Draw a line 6 inches long, as nearly as you can. Measure the line, and tell me how long it really is.

Draw a square, one inch on each side.

Draw a square with its sides 2 inches long, and divide it into four smaller squares.

How many square inches in a square, the sides of which are 2 inches long?

How many square inches in a square, the sides of which are 3 inches long?

Draw a square with its sides 3 inches long, and divide it into *nine* smaller squares.

NOTE. The Teacher should give exercises in measuring daily.

FOURTEEN. 14.

(a)  (b)

Look at the number picture marked (*a*).

How many dots are there in each row?

How many rows are there?

How many dots in the two rows together?

How many dots, then, are 2 times 7 dots?

How many columns are there of 2 dots each?

How many dots in the seven columns?

How many dots, then, are 7 times 2 dots?

If you divide 14 dots into two equal numbers, how many will there be in each number?

$14 \div 2 = ?$ $\frac{1}{2}$ of $14 = ?$ $2 \times 7 = ?$ $7 \times 2 = ?$

Count by 2's to 14.   How many 7's in 14?

How many skates are 7 pairs of skates?

Alice has 7 two-cent pieces. How many apples at one cent each can she buy?

How many weeks do 14 days make?

$2 \times 2 = ?$	$2 \times 5 = ?$	$4 \div 2 = ?$	$10 \div 5 = ?$
$2 \times 3 = ?$	$2 \times 6 = ?$	$6 \div 3 = ?$	$12 \div 2 = ?$
$2 \times 4 = ?$	$2 \times 7 = ?$	$8 \div 2 = ?$	$14 \div 7 = ?$
$8 + 4 = ?$	$6 + 5 = ?$	$9 + 5 = ?$	$8 + 9 = ?$
$9 + 6 = ?$	$7 + 6 = ?$	$5 + 7 = ?$	$8 + 5 = ?$
$7 + 8 = ?$	$9 + 4 = ?$	$6 + 9 = ?$	$7 + 9 = ?$
$14 - 8 = ?$	$17 - 8 = ?$	$14 - 5 = ?$	$16 - 7 = ?$
$15 - 7 = ?$	$14 - 9 = ?$	$13 - 6 = ?$	$13 - 7 = ?$
$16 - 9 = ?$	$13 - 5 = ?$	$12 - 7 = ?$	$18 - 9 = ?$

FIFTEEN. 15.

Look at the number picture marked (*a*).

How many dots are there in each row?

How many rows of dots are there?

How many dots in the three rows?

How many dots, then, are 3 times 5 dots?

How many dots are there in each column of dots?

How many columns of dots are there?

How many dots in the five columns?

How many dots, then, are 5 times 3 dots?

Look at the number picture marked (*b*).

How many sets of 5 each in 15?

Count by 3's to 15. Count by 5's to 15.

$15 \div 5 = ?$ $15 \div 3 = ?$ $\frac{1}{3}$ of $15 = ?$ $\frac{1}{5}$ of $15 = ?$

If one orange cost 3 cents, how many cents will 5 oranges cost? will 4 oranges cost?

How many pencils at a cent each can you buy with 3 five-cent pieces? with 2 five-cent pieces?

Find $\frac{1}{3}$ of 15 oranges; $\frac{1}{5}$ of 15 oranges.

Emily has 15 cents in five-cent pieces. How many five-cent pieces has she?

How many feet long is a string that is 5 yards long? 4 yards long? 3 yards long? 2 yards long?

What part of 15 pears are 5 pears? 3 pears?

How many inches are there in 1 foot and $\frac{1}{4}$ of a foot? in 1 foot and $\frac{1}{6}$ of a foot?

SIXTEEN. 16.

Look at the number picture marked (*a*).
How many dots are there in each row?
How many rows are there?
How many dots in the four rows?
How many dots, then, are 4 times 4 dots?

Look at the number picture marked (*b*).
How many dots are there in each row?
How many rows are there?
How many dots in the two rows?
How many dots, then, are 2 times 8 dots?
How many columns of 2 dots each are there?
How many dots in the eight columns?
How many dots, then, are 8 times 2 dots?
Count by 2's to 16.   Count by 4's to 16.
How many 2's in 16?   How many 4's in 16?

$4 \times 4 = ?$ $2 \times 8 = ?$ $8 \times 2 = ?$ $16 \div 4 = ?$
$16 \div 2 = ?$ $16 \div 8 = ?$ $15 \div 3 = ?$ $15 \div 5 = ?$
$\frac{1}{2}$ of $16 = ?$ $\frac{1}{8}$ of $16 = ?$ $\frac{1}{4}$ of $16 = ?$ $\frac{1}{5}$ of $15 = ?$

At 4 cents a quart, how many quarts of milk can you buy for 16 cents?

At 2 cents a pint, how many pints of milk can you buy for 16 cents?

At 8 cents a quart, how many quarts of berries can you buy for 16 cents?

OUNCES IN A POUND.

How many ounces make a pound?

Sixteen ounces make a pound.

How many ounces in ½ of a pound?

How many ounces in ¼ of a pound?

What part of a pound are 8 ounces?

What part of a pound are 4 ounces?

How many ounces in a quarter of a pound of tea?

How many ounces in a half of a pound of tea?

What will a pound of prunes cost, if half of a pound costs 8 cents?

What will a pound of raisins cost, if a quarter of a pound costs 4 cents?

If I buy three-quarters of a pound of candy, how many ounces of candy do I buy?

How many 4-ounce weights are equal to a pound weight? How many 8-ounce weights? How many 2-ounce weights? How many 1-ounce weights?

What part of a pound are 2 ounces? 4 ounces?

How many 1-ounce weights are equal to a 2-ounce weight? a 4-ounce weight? an 8-ounce weight?

If 1 egg weighs 2 ounces, how many eggs will it take to weigh a pound? a half-pound?

EIGHTEEN. 18.

(a)   (b)

How many dots in each row of dots marked (*a*)?

How many rows are there?

How many dots in the three rows?

How many dots, then, are 3 times 6 dots?

How many columns of dots are there?

How many dots in each column?

How many dots in the six columns?

How many dots, then, are 6 times 3 dots?

How many 6's in 18?   How many 3's in 18?

Look at the dots marked (*b*).

How many dots in the top row? in the bottom row? in the two rows?

How many dots, then, are 2 times 9 dots?

How many columns of 2 dots each are there?

How many dots, then, are 9 times 2 dots?

How many 2's in 18?   How many 9's in 18?

Count by 2's to 18.   2+2+2+2+2+2+2+2+2.

Count by 3's to 18.   3 + 3 + 3 + 3 + 3 + 3.

Count by 6's to 18.   6 + 6 + 6.

$2 \times 4 = ?$	$2 \times 5 = ?$	$2 \times 6 = ?$	$2 \times 7 = ?$
$2 \times 8 = ?$	$2 \times 9 = ?$	$18 \div 2 = ?$	$18 \div 3 = ?$
$18 \div 6 = ?$	$18 \div 9 = ?$	$9 + 9 = ?$	$18 - 9 = ?$
$\frac{1}{2}$ of $18 = ?$	$\frac{1}{3}$ of $18 = ?$	$\frac{1}{6}$ of $18 = ?$	$\frac{1}{9}$ of $18 = ?$

What part of 18 is 9?   What part of 18 is 6?

What part of 18 is 3?   What part of 18 is 2?

TWENTY. 20.

How many dots in each row of dots marked (a)?
How many rows are there ?
How many dots in the four rows ?
How many dots, then, are 4 times 5 dots ?
How many columns of dots are there ?
How many dots in each column ?
How many dots in the five columns ?
How many dots, then, are 5 times 4 dots ?
How many 5's in 20 ?   How many 4's in 20 ?
Count by 4's to 20.   Count by 5's to 20.

Look at the number picture marked (b).
How many dots in the top row ?
How many dots in the bottom row ?
How many dots in the two rows ?
How many dots, then, are 2 times 10 dots ?
How many columns of 2 dots each are there ?
How many dots in the 10 columns ?
How many dots, then, are 10 times 2 dots ?
How many 10's in 20 ?   How many 2's in 20 ?

$$4 \times 5 = ? \quad 5 \times 4 = ? \quad 2 \times 10 = ? \quad 10 \times 2 = ?$$
$$20 \div 4 = ? \quad 20 \div 5 = ? \quad 20 \div 2 = ? \quad 20 \div 10 = ?$$

Count by 2's to 19, beginning 1, 3, 5, etc.
Count by 3's to 19, beginning 1, 4, 7, etc.
Count by 3's to 20, beginning 2, 5, 8, etc.

ADDITION TABLE.

1	1	1	1	1	1	1	1	1	1
0	1	2	3	4	5	6	7	8	9
—	—	—	—	—	—	—	—	—	—
2	2	2	2	2	2	2	2	2	2
0	1	2	3	4	5	6	7	8	9
—	—	—	—	—	—	—	—	—	—
3	3	3	3	3	3	3	3	3	3
0	1	2	3	4	5	6	7	8	9
—	—	—	—	—	—	—	—	—	—
4	4	4	4	4	4	4	4	4	4
0	1	2	3	4	5	6	7	8	9
—	—	—	—	—	—	—	—	—	—
5	5	5	5	5	5	5	5	5	5
0	1	2	3	4	5	6	7	8	9
—	—	—	—	—	—	—	—	—	—
6	6	6	6	6	6	6	6	6	6
0	1	2	3	4	5	6	7	8	9
—	—	—	—	—	—	—	—	—	—
7	7	7	7	7	7	7	7	7	7
0	1	2	3	4	5	6	7	8	9
—	—	—	—	—	—	—	—	—	—
8	8	8	8	8	8	8	8	8	8
0	1	2	3	4	5	6	7	8	9
—	—	—	—	—	—	—	—	—	—
9	9	9	9	9	9	9	9	9	9
0	1	2	3	4	5	6	7	8	9
—	—	—	—	—	—	—	—	—	—

NOTE. The Teacher should copy this **addition table** on the board, and require *each pupil in turn* to name the sums as she touches the examples at random with a pointer. She should continue the drill daily until every pupil is absolutely certain of the required answer.

James had 2 peaches, and Tom had 5 peaches. How many did they have together?

Harry bought a quart of peanuts for 6 cents, and a lead pencil for 2 cents. How much money did he spend?

There are 7 apples on one limb, and 2 apples on another. How many apples on both limbs?

Susie has 8 white roses, and Alice has 2 red roses. How many roses have they together?

Nine boys are at play, and 2 boys are looking on. How many boys in all?

John had 10 marbles, and found 2 more. How many had he then?

Mary had 7 cherries, and her brother gave her 3 more. How many had she then?

Alice had 8 white chickens, and 3 brown chickens. How many chickens had she in all?

A farmer sold 9 bushels of corn at one time, and 3 bushels at another time. How many bushels did he sell in all?

Harry saw 5 birds sitting on a fence, and 4 birds on the ground. How many birds did he see?

Nora bought a quart of peanuts for 6 cents, and an orange for 4 cents. How much money did she spend?

A boy paid 3 cents for an orange, and 8 cents for some bananas. How much did he pay in all?

The cook used 6 eggs for a pudding, and 7 eggs for cake. How many eggs did she use?

Kate bought half a quire of note paper for 6 cents, and a bunch of envelopes for 5 cents. How much did she pay for the paper and envelopes?

Ernest found 7 eggs in one nest, and 4 eggs in another nest. How many eggs did he find in all?

Emma picked 8 quarts of berries, and Frank 4 quarts. How many quarts did they both pick?

There are 9 apples in one dish, and 4 apples in another. How many apples are there in all?

A farmer had 9 red cows, and 5 red and white cows. How many cows had he?

A boy rode 8 miles, and walked 5 miles. How many miles did he go?

A man paid 7 dollars for a ton of coal, and 5 dollars for a cord of wood. How many dollars did he pay for the coal and wood together?

A man worked 6 days one week, and 5 days the next week. How many days did he work in all?

School begins at 9 o'clock in the morning, and continues 3 hours. What o'clock is it when school is dismissed?

John had 6 cents, and earned 6 cents more. How much money had he then?

Jane paid 9 cents for a slate, and 4 cents for some paper. How much did the slate and paper together cost?

Olive paid 5 cents for a spool of silk, and 9 cents for two yards of ribbon. How much did she pay in all?

A farmer sold 7 lambs at one time, and 6 lambs at another time. How many lambs did he sell?

A milkman has 8 Dutch cows, and 6 Durham cows. How many cows has he?

James paid 9 dollars for a coat, and 6 dollars for a vest. How much did he pay for both?

In a school one class has 9 girls, and another has 7 girls. How many girls have the two classes?

Harry caught 8 trout, and Tom caught 7 trout. How many did they catch in all?

There are 8 yards of ribbon in one roll, and 9 yards in another. How many yards are there in the two rolls?

In a game of baseball 9 persons play on one side, and 9 persons on the other side. How many persons does it take to play the game?

If a boy buys an orange for 4 cents, a pear for 3 cents, and an apple for 2 cents, how much does he pay for all?

Alice bought a postage stamp for 5 cents, another for 4 cents, and another for 3 cents. How much did she pay for the three stamps?

A 5-cent piece, a 2-cent piece, and 9 single cents are equal to how many cents?

Tom paid 3 cents for a top, 2 cents for a ball, and 7 cents for a book. How much did he pay?

James hoed 4 rows of potatoes, George hoed 5 rows, and Oscar hoed 6 rows. How many rows did they all hoe?

There are 7 pies on one shelf, and 9 pies on another shelf. How many pies are there on the two shelves?

There are 6 eggs in one nest, 4 in another, and 3 in another. How many eggs are there in the three nests together?

Nora paid 9 cents for ribbon, 5 cents for buttons, and 3 cents for pins. How much did she pay for all?

A lady bought a dress for 9 dollars, a hat for 4 dollars, and a parasol for 3 dollars. How much did she pay for all?

If a table is 8 feet long and 5 feet wide, what is the number of feet in one side and the two ends?

Harry had 7 five-cent pieces, 4 two-cent pieces, and 8 one-cent pieces. How many pieces of money had he in all?

Alice picked 6 quarts of berries, Kate 6 quarts, and Florence 5 quarts. How many quarts did they all pick together?

In a certain garden there are 6 pear trees, 7 peach trees, and 5 cherry trees. How many trees are there in the garden?

James picked 7 boxes of strawberries on Monday, 3 boxes on Tuesday, and 6 boxes on Wednesday. How many boxes did he pick?

Frank bought a pencil for 4 cents, a penholder for 3 cents, and a block of paper for 9 cents. How much did he have to pay for all?

SUBTRACTION TABLE.

1	2	3	4	5	6	7	8	9	10
−1	−1	−1	−1	−1	−1	−1	−1	−1	−1
2	3	4	5	6	7	8	9	10	11
−2	−2	−2	−2	−2	−2	−2	−2	−1	−2
3	4	5	6	7	8	9	10	11	12
−3	−3	−3	−3	−3	−3	−3	−3	−3	−3
4	5	6	7	8	9	10	11	12	13
−4	−4	−4	−4	−5	−5	−5	−5	−5	−5
5	6	7	8	9	10	11	12	13	14
−5	−5	−5	−5	−5	−5	−5	−5	−5	−5
6	7	8	9	10	11	12	13	14	15
−6	−6	−6	−6	−6	−6	−6	−6	−6	−6
7	8	9	10	11	12	13	14	15	16
−7	−7	−7	−7	−7	−7	−7	−7	−7	−7
8	9	10	11	12	13	14	15	16	17
−8	−8	−8	−8	−8	−8	−8	−8	−8	−8
9	10	11	12	13	14	15	16	17	18
−9	−9	−9	−9	−9	−9	−9	−9	−9	−9
10	11	12	13	14	15	16	17	18	19
−10	−10	−10	−10	−10	−10	−10	−10	−10	−10

NOTE. The Teacher should copy this **subtraction table** on the board, and require *each pupil in turn* to name the differences as she touches the examples at random with a pointer. She should continue the drill daily until every pupil is absolutely certain of the required answer.

Robert had 9 cents, and spent 4 of them for an orange.   How many cents had he left?

From 'a string 14 inches long, 4 inches were cut off.   How many inches remained?

A hen had 11 chickens.   A hawk caught 4 of them.   How many chickens were left?

There were 13 crows on the ground.   Four of them flew away.   How many were left?

A baker sold 10 loaves of bread in the morning, and 7 loaves in the afternoon.   How many more loaves did he sell in the morning than in the afternoon?

A farmer raised 19 barrels of apples, and sold 9 barrels?   How many barrels had he left?

From a dozen cans of tomatoes three cans were used.   How many cans were left?

Alice has 7 dolls.   How many more must she get in order to have 10 dolls?

Ernest has 14 ducks and 6 geese.   How many more ducks has he than geese?

Ellen is 11 years old, and Susan is 7 years old. How many years older is Ellen than Susan?

A carpenter had a board 15 feet long.   He sawed off a piece 7 feet long.   How many feet long was the other piece?

A farmer sold a calf for 11 dollars, and a pig for 3 dollars.   How much more did he receive for the calf than for the pig?   How much did he receive for the calf and pig together?

A milkman has 13 cows. Six of them are dark red cows, and the rest are black and white. How many of them are black and white?

Robert has to travel 16 miles. How many miles remain after he has gone 7 miles?

How many eggs must you put with 7 eggs in order to have a dozen eggs?

There were 12 rats in the stable, but 5 of them were caught in a trap. How many rats escaped?

In a pigeon house there were 16 pigeons, but 9 flew away. How many pigeons remained?

Mary hemmed 15 handkerchiefs, and Ellen only 7. How many more did Mary hem than Ellen?

A farmer had 17 lambs. He sold 8 of them. How many lambs had he left?

A sitting hen had 13 eggs under her, but only 9 chickens came out. How many eggs had no chickens?

Seventeen spiders waited for flies, but 7 spiders waited without catching any. How many spiders caught flies?

John had 13 cents, and paid 5 cents for car fare. How many cents had he left?

George earned 13 cents, and spent 8 cents. How many cents had he left?

Florence had 16 pinks. Eight were red, and the rest were white. How many were white?

Bertha had 18 chickens. Nine were white, and the rest were black. How many were black?

Peter raised 13 melons, and sold 9 of them. How many were left?

There were 15 eggs in a nest, but 9 of them were carried into the house. How many were left?

Fourteen lilies were growing in a field, but a boy picked 9 of them. How many lilies were left?

Harry has earned 9 cents by selling newspapers. How many more cents must he earn in order to have 16 cents?

Robert had 17 rows of peas. He has hoed 9 rows. How many more rows has he to hoe?

From a dozen cans of peaches 9 cans were used. How many cans were left?

A gardener raised 11 dozen cabbages, and sold 9 dozen. How many dozen had he left?

A farmer had 10 oxen, but he sold one pair of them. How many oxen had he left?

A butter dealer had 11 pounds of butter, and sold 8 pounds. How many pounds were left?

A tea merchant bought 14 chests of tea. When he had sold 8 chests, how many had he left?

From 14 yards of cloth a merchant sold 5 yards. How many yards were left?

Richard had 10 lambs. He sold 3 of his lambs to one man, and 2 to another man. How many lambs remained?

Florence had 15 roses. Three of the roses were yellow, three were white, and the rest were red. How many red roses did she have?

TENS.

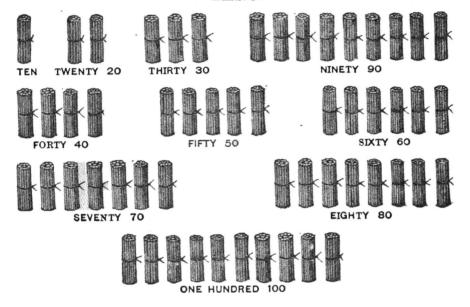

TEN TWENTY 20 THIRTY 30 NINETY 90 FORTY 40 FIFTY 50 SIXTY 60 SEVENTY 70 EIGHTY 80 ONE HUNDRED 100

What do we call 2 tens? 3 tens? 4 tens? 5 tens? 6 tens? 7 tens? 8 tens? 9 tens? 10 tens?

How many tens make ninety? thirty? one hundred? seventy? fifty? forty? sixty? eighty?

If I pay 6 ten-cent pieces for peaches, and 3 ten-cent pieces for pears, how many cents do I spend?

If I have 6 ten-cent pieces in one pocket, and 4 in another, how much money have I?

How many tens are 3 tens and 4 tens? 5 tens and 2 tens? 4 tens and 4 tens? 5 tens and 5 tens?

How many ten-cent pieces make a dollar?

Twenty is sometimes called a score.

How many years are 2 score years?

How old is a man who is 4 score years old?

How many years are 3 score and ten years?

Copy, and write the results:

20	20	50	70	30	60	50
+50	+60	+30	+20	+40	+30	+40

70	80	40	50	60	80	90
+30	+10	+40	+50	+40	+20	+10

50	60	70	80	90	30	50
−20	−30	−10	−50	−20	−20	−30

70	90	80	70	90	80	40
−30	−70	−40	−20	−30	−60	−20

$3 \times 20 = ?$	$2 \times 20 = ?$	$5 \times 10 = ?$
$4 \times 20 = ?$	$5 \times 20 = ?$	$5 \times 20 = ?$
$3 \times 30 = ?$	$2 \times 30 = ?$	$9 \times 10 = ?$
$4 \times 10 = ?$	$2 \times 40 = ?$	$10 \times 10 = ?$

$80 \div 4 = ?$	$60 \div 2 = ?$	$\frac{1}{2}$ of $40 = ?$
$20 \div 2 = ?$	$40 \div 4 = ?$	$\frac{1}{3}$ of $60 = ?$
$60 \div 3 = ?$	$80 \div 10 = ?$	$\frac{1}{4}$ of $80 = ?$
$90 \div 3 = ?$	$100 \div 10 = ?$	$\frac{1}{5}$ of $50 = ?$

You have already learned that we write the figure for the number of tens in the *second* place from the right. In what place, counting from the right, do we write the *hundreds* of a number?

Write on the board the number that contains six hundreds, no tens, and five ones.

If you rub out the 0, what does the number become?

Two tens and one   make twenty-one,   **21.**
Two tens and two   make twenty-two,   **22.**
Two tens and three make twenty-three, **23.**
Two tens and four   make twenty-four, **24.**
Two tens and five   make twenty-five, **25.**
Two tens and six   make twenty-six,   **26.**
Two tens and seven make twenty-seven, **27.**
Two tens and eight make twenty-eight, **28.**
Two tens and nine   make twenty-nine, **29.**

What are the names of the numbers made up of 3 tens and 1? 3 tens and 2? 3 tens and 3? 3 tens and 4? 3 tens and 5? 3 tens and 6? 3 tens and 7? 3 tens and 8? 3 tens and 9?

What are the names of the numbers made up of 4 tens and 1? 4 tens and 2? 4 tens and 3? 4 tens and 4? 4 tens and 5? 4 tens and 6? 4 tens and 7? 4 tens and 8? 4 tens and 9?

What are the names of the numbers made up of 5 tens and 1? 5 tens and 2? 5 tens and 3? 5 tens and 4? 5 tens and 5? 5 tens and 6? 5 tens and 7? 5 tens and 8? 5 tens and 9?

What are the names of the numbers made up of 6 tens and 1? 6 tens and 2? 6 tens and 3? 6 tens and 4? 6 tens and 5? 6 tens and 6? 6 tens and 7? 6 tens and 8? 6 tens and 9?

What are the names of the numbers made up of 7 tens and 1? 7 tens and 2? 7 tens and 3? 7 tens and 4? 7 tens and 5? 7 tens and 6? 7 tens and 7?

Read the numbers: 78; 79; 81; 82; 83; 84; 85; 86; 87; 88; 89.

Read the numbers: 91; 92; 93; 94; 95; 96; 97; 98; 99; 100; 200; 300; 400.

How many more tens has the number 84 than 72? 63 than 31? 55 than 15? 42 than 2? 95 than 80? 65 than 50? 94 than 43? 99 than 39?

Copy, and complete:

$18 = 10 + ?$	$26 = 2 \times 10 + ?$	$67 = 6 \times 10 + ?$
$14 = 10 + ?$	$37 = 3 \times 10 + ?$	$84 = 8 \times 10 + ?$
$13 = 10 + ?$	$24 = 2 \times 10 + ?$	$85 = 8 \times 10 + ?$
$19 = 10 + ?$	$35 = 3 \times 10 + ?$	$89 = 8 \times 10 + ?$
$12 = 10 + ?$	$39 = 3 \times 10 + ?$	$86 = 8 \times 10 + ?$
$15 = 10 + ?$	$41 = 4 \times 10 + ?$	$88 = 8 \times 10 + ?$
$16 = 10 + ?$	$47 = 4 \times 10 + ?$	$95 = 9 \times 10 + ?$
$17 = 10 + ?$	$43 = 4 \times 10 + ?$	$97 = 9 \times 10 + ?$
$11 = 10 + ?$	$55 = 5 \times 10 + ?$	$93 = 9 \times 10 + ?$
$20 = 10 + ?$	$59 = 5 \times 10 + ?$	$96 = 9 \times 10 + ?$
$50 = 10 + ?$	$51 = 5 \times 10 + ?$	$98 = 9 \times 10 + ?$
$70 = 10 + ?$	$52 = 5 \times 10 + ?$	$99 = 9 \times 10 + ?$

Copy, and add:

5	6	2	9	7	5	2	4	6	9
3	3	5	2	5	7	5	6	2	5
8	7	9	4	3	4	7	6	6	4

7	8	5	4	5	6	3	5	4
7	2	6	3	2	1	3	6	7
4	9	6	5	7	4	4	5	6

Copy, and add, adding the *ones* first:

22	31	33	25	18	35
21	33	11	21	30	11
23	12	13	11	21	21
32	23	31	22	20	22

34	60	40	41	36	23
12	17	25	34	21	22
30	12	13	13	20	32
13	10	11	10	12	11

Copy, and subtract, subtracting the *ones* first:

65	87	98	78	63	77
− 43	− 55	− 67	− 52	− 51	− 35

99	76	95	46	37	89
− 44	− 66	− 54	− 22	− 21	− 65

62	71	92	85	74	52
− 40	− 50	− 70	− 30	− 43	− 50

Copy, and multiply, multiplying the *ones* first:

21	32	13	24	34	42
2	2	2	2	2	2

31	23	33	43	44	30
2	2	2	2	2	2

11	10	23	12	32	33
3	3	3	3	3	3

10	11	12	20	21	22
4	4	4	4	4	4

TWENTY-ONE. 21.

How many dots in each row of dots marked (*a*)?
How many rows of dots?
How many dots in the three rows together?
How many dots, then, are 3 times 7 dots?
How many dots in each column of dots?
How many columns of dots?
How many dots in the seven columns?
How many dots, then, are 7 times 3 dots?
How many 3's in 21?

Look at the number picture marked (*b*).
How many 7's in 21?

$3 \times 7 = ?$	$21 \div 3 = ?$	$\frac{1}{3}$ of $21 = ?$
$7 \times 3 = ?$	$21 \div 7 = ?$	$\frac{1}{7}$ of $21 = ?$

If a pair of boots costs 7 dollars, what will 3 pairs of boots cost? 2 pairs of boots?

If an orange costs 3 cents, what will 7 oranges cost? 6 oranges? 5 oranges? 4 oranges?

Divide 21 oranges equally among 3 boys. How many oranges will each boy have?

Divide 21 oranges equally among 7 boys. How many oranges will each boy have?

There are 21 apples in a basket, and James takes one-third of them. How many apples does he take?

If he had taken $\frac{1}{7}$ of them, how many would he have taken?

TWENTY-FOUR. 24.

(a) (b)

How many dots in each row of dots marked (a)?
How many rows of dots?
How many dots in the three rows?
How many dots, then, are 3 times 8 dots?
How many dots in each column of dots?
How many columns are there?
How many dots in the eight columns?
How many dots, then, are 8 times 3 dots?

Look at the dots marked (b).
How many dots in each row?
How many rows of dots?
How many dots in the four rows?
How many dots, then, are 4 times 6 dots?
How many dots in each column of dots?
How many columns are there?
How many dots in the six columns?
How many dots, then, are 6 times 4 dots?

$3 \times 8 = ?$ $8 \times 3 = ?$ $4 \times 6 = ?$ $6 \times 4 = ?$

How many 3's in 24? How many 8's? How many 4's? How many 6's?

$24 \div 3 = ?$ $24 \div 4 = ?$ $24 \div 6 = ?$ $24 \div 8 = ?$
$\frac{1}{3}$ of 24 = ? $\frac{1}{4}$ of 24 = ? $\frac{1}{6}$ of 24 = ? $\frac{1}{8}$ of 24 = ?

$4 \times 2 = ?$ $4 \times 3 = ?$ $4 \times 4 = ?$ $4 \times 5 = ?$ $4 \times 6 = ?$
$5 \times 2 = ?$ $5 \times 3 = ?$ $5 \times 4 = ?$ $6 \times 3 = ?$ $6 \times 4 = ?$

TWENTY-FIVE. 25.

How many dots in each row of dots marked (*a*)?

How many rows are there?

How many dots in the five rows?

How many dots, then, are 5 times 5 dots?

How many 5's in 25? $\frac{1}{5}$ of 25 = ?   25 ÷ 5 = ?

Count by 5's to 25?   Count by 4's to 24.

NOTE. Assist the pupil by dots to count by 3's, 4's, etc., but only so long as such assistance is necessary.

Count by 3's to 24.   Count by 2's to 24.

Count by 6's to 24.   Count by 8's to 24.

Count by 3's to 25, beginning 1, 4, etc.

Count by 3's to 23, beginning 2, 5, etc.

Count by 4's to 25, beginning 1, 5, etc.

Count by 4's to 22, beginning 2, 6, etc.

Count by 4's to 23, beginning 3, 7, etc.

There are 5 plates in a row, and each plate has 5 apples on it. How many apples on the 5 plates?

If you divide 25 oranges equally among five little girls, how many oranges will each girl have?

If you have 25 oranges, how many times can you give away oranges if you give 5 each time?

How many eggs make a dozen? a half-dozen?

How many inches make a foot? How many feet a yard? How many quarts a gallon?

TWENTY-SEVEN. 27.

How many dots in each row of dots marked (a)?

How many rows are there?

How many dots in the three rows together?

How many dots, then, are 3 times 9 dots?

How many dots in each column of dots?

How many columns are there?

How many dots in the nine columns?

How many dots, then, are 9 times 3 dots?

How many 3's in 27? How many 9's?

$27 \div 3 = ?$ $\frac{1}{3}$ of $27 = ?$ $27 \div 9 = ?$ $\frac{1}{9}$ of $27 = ?$

How many three-cent stamps can I buy for 27 cents? for 24 cents? for 21 cents?

In one yard there are 3 feet. How many feet in 9 yards? in 8 yards? in 7 yards? in 6 yards?

At 9 cents a quart, how much will 3 quarts of berries cost? 2 quarts of berries?

$2 \times 1 = ?$	$3 \times 1 = ?$	$4 \times 2 = ?$	$6 \times 2 = ?$
$2 \times 2 = ?$	$3 \times 2 = ?$	$4 \times 3 = ?$	$6 \times 3 = ?$
$2 \times 3 = ?$	$3 \times 3 = ?$	$4 \times 4 = ?$	$6 \times 4 = ?$
$2 \times 4 = ?$	$3 \times 4 = ?$	$4 \times 5 = ?$	$7 \times 2 = ?$
$2 \times 5 = ?$	$3 \times 5 = ?$	$4 \times 6 = ?$	$7 \times 3 = ?$
$2 \times 6 = ?$	$3 \times 6 = ?$	$5 \times 2 = ?$	$8 \times 2 = ?$
$2 \times 7 = ?$	$3 \times 7 = ?$	$5 \times 3 = ?$	$8 \times 3 = ?$
$2 \times 8 = ?$	$3 \times 8 = ?$	$5 \times 4 = ?$	$9 \times 2 = ?$
$2 \times 9 = ?$	$3 \times 9 = ?$	$5 \times 5 = ?$	$9 \times 3 = ?$

TWENTY-EIGHT. 28.

How many dots in each row of dots marked (a)?

How many rows are there?

How many dots in the four rows together?

How many dots, then, are 4 times 7 dots?

How many dots in each column?

How many columns of dots are there?

How many dots in the seven columns?

How many dots, then, are 7 times 4 dots?

How many 4's in 28?   How many 7's in 28?

$4 \times 7 = ?$ $7 \times 4 = ?$ $28 \div 4 = ?$ $28 \div 7 = ?$

At 4 cents a quart, what will 6 quarts of milk cost? What will 7 quarts cost?

At 6 cents a quart, what will 4 quarts of berries cost? What will 3 quarts cost?

At 7 cents a quart, what will 4 quarts of berries cost? What will 3 quarts cost?

At 7 cents a cake, how many cakes of maple sugar can you buy for 28 cents?

If it takes 4 men 7 days to dig a certain ditch, how long will it take 1 man to dig the ditch?

If it takes a man 28 days to build a certain wall, how many days will it take him to build a quarter of the wall? Three-quarters of the wall?

What part of 28 is 7? What part of 24 is 4?
What part of 24 is 8? What part of 27 is 9?

THIRTY. 30.

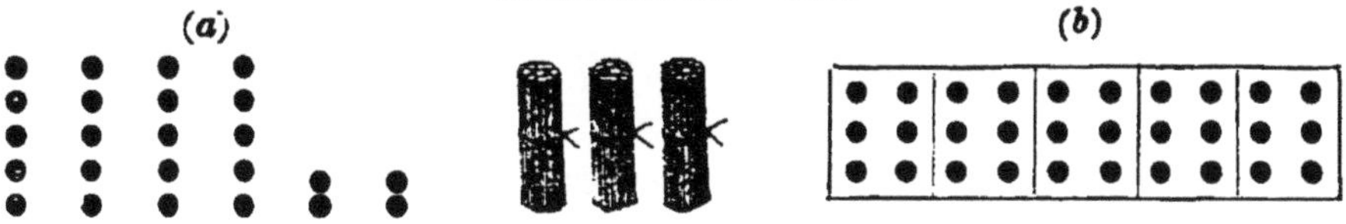

How many dots in each row of dots marked (*a*)?

How many rows are there?

How many dots in the five rows?

How many dots, then, are 5 times 6 dots?

How many dots in each column of dots?

How many columns of dots are there?

How many dots in the six columns?

How many dots, then, are 6 times 5 dots?

How many 6's in 30? How many 5's in 30?

What part of 30 is 6? What part of 30 is 5?

$5 \times 6 = ?$ $6 \times 5 = ?$ $30 \div 5 = ?$ $30 \div 6 = ?$

How many cents are 6 five-cent pieces?

How many five-cent stamps can you buy for 30 cents? for 25 cents? for 20 cents?

When berries are 6 cents a quart, how many quarts can you buy for 30 cents? for 24 cents?

How many more is $\frac{1}{5}$ of 30 than $\frac{1}{6}$ of 30?

How many tens in 30? How many fives in $\frac{1}{6}$ of 30? in $\frac{1}{3}$ of 30? in $\frac{1}{2}$ of 30?

How many *sixths* of 30 must you take to have $\frac{1}{3}$ of 30? to have $\frac{1}{2}$ of 30?

How many *sixths* of any number must you take to have $\frac{1}{3}$ of the number? to have $\frac{1}{2}$ of the number?

How many inches in $\frac{1}{3}$ of a foot? in $\frac{2}{6}$ of a foot?

How many inches in $\frac{1}{2}$ of a foot? in $\frac{3}{6}$ of a foot?

SLATE ADDITION.

Add 8 ones to 6 tens and 7 ones.

Write the 6 tens and 7 ones 67
Then write the 8 ones under the 7 ones . . .   8
Add the ones.   75
How many are 8 ones and 7 ones?   15.
How many tens and how many ones in 15?
Write the 5 ones in the **ones'** place under 8.
What shall be done with the 1 ten?
Add it to the 6 tens, and we have 7 tens.
Now write the 7 tens in the **tens'** place.
Read the answer. How many tens and ones in 75?

Add 7 ones to 2 tens and 8 ones.
Add 8 ones to 2 tens and 5 ones.
Add 6 ones to 3 tens and 7 ones.
Add 4 ones to 4 tens and 9 ones.
Add 9 ones to 4 tens and 7 ones.
Add 5 ones to 5 tens and 5 ones.
Add 7 ones to 7 tens and 3 ones.

| 68 | 76 | 47 | 28 | 35 | 24 |
| +8 | +5 | +6 | +4 | +8 | +9 |

| 56 | 57 | 65 | 69 | 63 | 55 |
| +4 | +9 | +7 | +4 | +8 | +5 |

| 84 | 87 | 88 | 79 | 88 | 79 |
| +7 | +3 | +9 | +6 | +8 | +9 |

| 33 | 46 | 57 | 64 | 77 | 86 |
| +7 | +8 | +7 | +9 | +9 | +9 |

Add 3 tens and 7 ones to 4 tens and 6 ones.

Write the 4 tens and 6 ones 46
Then the 3 tens and 7 ones 37
Add the ones.   83

How many are 7 ones and 6 ones?   13.

How many tens and how many ones in 13?

Write the 3 ones in the **ones'** place under the 7.

What shall be done with the 1 ten in 13?

Add it to the tens.

1 ten and 3 tens are? and 4 tens more?

Write the 8 in the tens' place.

Therefore the **sum** of 46 and 37 is 83.

Add 5 tens and 3 ones to 1 ten and 8 ones.

Add 7 tens and 6 ones to 1 ten and 5 ones.

Add 3 tens and 7 ones to 3 tens and 6 ones.

Add 3 tens and 3 ones to 3 tens and 9 ones.

Add 2 tens and 5 ones to 5 tens and 5 ones.

Add 4 tens and 9 ones to 4 tens and 8 ones.

Add 6 tens and 4 ones to 1 ten and 9 ones.

Add 3 tens and 8 ones to 4 tens and 7 ones.

64	48	76	57	35	56
18	29	18	19	56	24
55	28	55	35	68	39
38	36	29	16	19	26
48	65	53	57	48	34
32	19	28	35	27	28

Add :

67	74	57	29	39	59
19	16	38	34	47	38
36	19	32	17	23	18
14	46	28	23	19	57
20	18	47	27	30	35
14	36	15	22	17	17
17	23	18	25	49	24
34	49	56	38	28	18
16	24	26	27	34	57
12	15	16	27	39	19
25	23	39	47	39	35
28	28	14	22	23	39
27	35	27	17	26	14
38	19	38	39	26	25
28	57	25	12	19	37
35	18	28	14	17	14
16	18	29	38	45	26
18	19	25	34	39	19
27	24	36	24	12	45
27	37	56	17	19	28
28	19	12	19	29	38
35	25	22	44	39	18

THIRTY-TWO. 32.

(a)   (b)

How many dots in each row of dots marked (a)?

How many rows of dots?

How many dots in the 4 rows?

How many dots, then, are 4 times 8 dots?

How many dots in each column of dots?

How many columns of dots are there?

How many dots in the eight columns?

How many dots, then, are 8 times 4 dots?

$4 \times 8 = ?$ $8 \times 4 = ?$ $32 \div 4 = ?$ $32 \div 8 = ?$

How many shoes will a blacksmith need to shoe 8 horses all round?

A teamster has 32 horses. How many four-horse teams can he form? How many eight-horse teams?

At 4 cents a quart, how many quarts of milk can you buy for 32 cents? for 28 cents?

At 8 cents a pint, how many pints of cream can you buy for 32 cents? for 24 cents?

Four weeks make a lunar month. How many weeks are there in 8 lunar months? in 7? in 6?

At 8 cents a pound, how much will 4 pounds of sugar cost? 3 pounds? 2 pounds?

How many pears in ¼ of 32 pears? in ⅛ of 32 pears? in ¼ of 24 pears? in ⅔ of 24 pears?

THE PECK.

PINT.   QUART.   PECK.

NOTE. These wooden measures are used for measuring *dry* articles, such as oats, wheat, beans, potatoes, etc.

How many pints in one quart?

How many quarts make one peck?*

Eight quarts make one peck.

How many 2-quart measures of oats will a peek measure hold? How many 4-quart measures?

One quart of oats is what part of a peek of oats? Two quarts of oats are what part of a peek? four quarts?

How many quarts in 2 peeks? in 4 peeks?

If the peck measure is half-full of beans, how many more quarts of beans will it hold?

If the peek measure is a quarter-full of oats, how many more quarts will it hold?

If the peek measure is three-quarters full of cranberries, how many quarts of cranberries are in it? How many more quarts will it hold?

How many quarts in ½ of a peck? in ¼ of a peck? in ¾ of a peck?

At 2 cents a quart, what will a peek of corn cost?

At 3 cents a quart, what will a peek of nuts cost?

At 4 cents a quart, what will a peek of peas cost?

* Let the pupil discover the answer *by trial.*

THE BUSHEL.

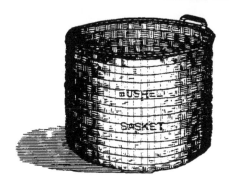 

How many pints make a quart?

How many quarts make a peck?

How many pecks make a bushel?

Four pecks make one bushel.

How many pecks in a half-bushel?

One peck of corn is what part of a bushel of corn?

Two pecks are what part of a bushel? Three pecks are what part of a bushel?

How many quarts in a peck of berries?

How many quarts in a half-bushel of berries?

How many quarts in a bushel of berries?

How many quarts in three-quarters of a bushel?

In 24 quarts how many pecks?

In 32 quarts how many pecks?

If a bushel basket is half-full of apples, how many more pecks of apples will it hold?

If a bushel basket is three-quarters full of apples, how many more pecks of apples will it hold?

A bushel of oats weighs 32 pounds. How much does a peck weigh? How much do 4 quarts weigh?

What part of a bushel are 4 quarts? 8 quarts?

THIRTY-FIVE. 35.

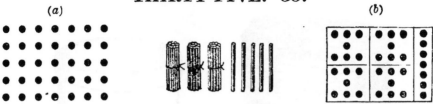

(a)   (b)

How many dots in each row of dots marked (a)?

How many rows of dots?

How many dots in the five rows?

How many dots, then, are 5 times 7 dots?

How many dots in each column of dots?

How many columns of dots?

How many dots in the seven columns?

How many dots, then, are 7 times 5 dots?

How many 7's in 35?   How many 5's in 35?

$5 \times 7 = ?$ $7 \times 5 = ?$ $35 \div 7 = ?$ $35 \div 5 = ?$

How many halves of a number make the entire number? How many thirds? How many fourths? How many fifths? How many sixths? How many sevenths?

$\frac{1}{2}$ of 10 = ?	$\frac{1}{3}$ of 12 = ?	$\frac{1}{4}$ of 20 = ?
$\frac{1}{2}$ of 12 = ?	$\frac{1}{3}$ of 15 = ?	$\frac{1}{4}$ of 24 = ?
$\frac{1}{2}$ of 14 = ?	$\frac{1}{3}$ of 18 = ?	$\frac{1}{5}$ of 25 = ?
$\frac{1}{2}$ of 16 = ?	$\frac{1}{3}$ of 21 = ?	$\frac{1}{5}$ of 35 = ?
$\frac{1}{2}$ of 18 = ?	$\frac{1}{3}$ of 24 = ?	$\frac{1}{7}$ of 35 = ?

At seven dollars a cord, how many cords of wood can be bought for 35 dollars? for 21 dollars?

At 5 cents a ride, how many street-car rides can be taken for 35 cents? for 25 cents? for 15 cents?

THIRTY-SIX. 36.

(a) (b)

How many dots in each row of dots marked (a)?

How many rows?

How many dots in the four rows?

How many dots, then, are 4 times 9 dots?

How many dots in each column of dots?

How many columns?

How many dots in the nine columns?

How many dots, then, are 9 times 4 dots?

How many dots in each row of dots marked (b)?

How many rows?

How many dots in the six rows?

How many dots, then, are 6 times 6 dots?

$4 \times 9 = ?$	$9 \times 4 = ?$	$6 \times 6 = ?$	$36 \div 4 = ?$
$36 \div 9 = ?$	$36 \div 6 = ?$	$\frac{1}{4}$ of $36 = ?$	$\frac{1}{6}$ of $36 = ?$

How many four-cent stamps can I buy for 36 cents? for 28 cents? for 32 cents? for 24 cents?

At 9 cents a yard, how many yards of calico can I buy for 36 cents? for 18 cents? for 27 cents?

At 6 cents a quart, how many quarts of milk can I buy for 36 cents? for 24 cents? for 30 cents?

$4 \times 2 = ?$	$4 \times 4 = ?$	$4 \times 6 = ?$	$4 \times 8 = ?$
$4 \times 3 = ?$	$4 \times 5 = ?$	$4 \times 7 = ?$	$4 \times 9 = ?$

FORTY. 40.

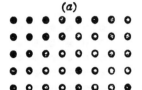

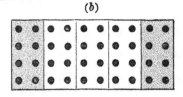

How many dots in each row of dots marked (*a*)?

How many rows?

How many dots in the five rows?

How many dots, then, are 5 times 8 dots?

How many dots in each column of dots?

How many columns?

How many dots in the eight columns?

How many dots, then, are 8 times 5 dots?

How many 5's in 40?   How many 8's in 40?

$5 \times 8 = ?$ $8 \times 5 = ?$ $40 \div 5 = ?$ $40 \div 8 = ?$

At 5 dollars a barrel, how many barrels of flour can you buy for 40 dollars? for 35 dollars?

At 8 cents a bottle, how many bottles of ink can you buy for 40 cents? for 32 cents?

If one loaf of bread is worth 5 cents, how many cents are 8 loaves worth? 6 loaves?

If a melon is worth 8 cents, how many cents are 5 melons worth? 4 melons?

How much will a boy earn in 9 weeks, if he earns 4 dollars a week?

$\frac{1}{5}$ of 20 = ?	$\frac{1}{2}$ of 16 = ?	$\frac{1}{8}$ of 16 = ?
$\frac{1}{5}$ of 30 = ?	$\frac{1}{4}$ of 32 = ?	$\frac{1}{8}$ of 32 = ?
$\frac{1}{5}$ of 40 = ?	$\frac{1}{4}$ of 40 = ?	$\frac{1}{8}$ of 40 = ?

SLATE SUBTRACTION.

The result obtained from subtracting a smaller number from a larger is called the **remainder** or **difference**. The smaller number is called the **subtrahend**; and the larger number, the **minuend**.

From 5 tens and 3 ones take 2 tens and 8 ones.

Write the 5 tens and 3 ones. 53
Write the 2 tens and 8 ones below 28
Draw a line underneath 25

We cannot take 8 ones from 3 ones. We therefore take 1 of the 5 tens and put with the 3 ones.

NOTE. Illustrate this. Let the pupil take a bundle of ten, and slipping off the rubber bands put the ten ones with the three ones.

We now have 13 ones, and 8 ones from 13 ones leave 5 ones. We write the 5 in the ones' place.

As we have taken 1 ten from the 5 tens, we have only 4 tens left, and 2 tens from 4 tens leave 2 tens.

We write the 2 in the tens' place, and have for the remainder 2 tens and 5 ones; that is, 25.

NOTE. The entire work may be shown as follows:

$$\begin{array}{ll} 53 & 40 + 13 \\ 28 & 20 + 8 \\ \hline 25 & 20 + 5 = 25. \end{array}$$

The pupils, however, must be taught from the first to do the work without any change of the figures.

75	23	33	31	37	86
−6	−4	−5	−3	−8	−7

67	35	37	32	46	82
−8	−9	−9	−7	−7	−3

Slate exercises:

75	42	33	64	83	92
− 37	− 25	− 16	− 28	− 38	− 29

50	41	42	56	35	52
− 29	− 24	− 15	− 27	− 26	− 28

48	42	62	55	61	72
− 19	− 29	− 33	− 27	− 37	− 36

70	52	85	75	85	60
− 37	− 39	− 16	− 36	− 28	− 48

98	96	73	86	83	57
− 69	− 27	− 57	− 69	− 27	− 18

74	67	85	91	80	61
− 37	− 19	− 38	− 64	− 55	− 28

64	73	81	80	43	82
− 45	− 26	− 33	− 43	− 26	− 57

94	72	91	80	51	90
− 18	− 19	− 29	− 37	− 22	− 23

87	95	93	90	73	83
− 19	− 26	− 38	− 43	− 37	− 35

Out of 16 eggs 7 were used for cooking. How many eggs were left?

In a class of 14 pupils there are 5 boys. How many girls are there in the class?

In a class of 13 pupils there are 6 girls. How many boys are there in the class?

Out of 15 signal flags, 8 are white, and the rest blue. How many flags are blue?

One package of tea weighs 16 ounces, and another weighs 8 ounces. How many more ounces in one package than in the other?

How much deeper is a well 21 feet deep than a well 18 feet deep?

How many more are 13 ducks than 9 ducks?

A man has 17 miles to go. After he has gone 9 miles, how many more has he to go?

From a board 16 inches long, a piece 9 inches long was cut off. How many inches long was the other piece?

A farmer had 13 lambs and sold 5 of them. How many had he left?

In a brood of 14 chickens 6 are white, and the rest brown. How many chickens are brown?

There were 13 crows on the ground. 7 flew away. How many were left on the ground?

What number must you add to 9 to get 12?

What number must you add to 3 to get 11?

What number must you take from 11 to get 5?

What number must you take from 14 to get 8?

The number 259 is read *two hundred fifty-nine*, and is composed of 2 hundreds, 5 tens, and 9 ones.

Read, and give the number of hundreds, of tens, and of ones, in the following numbers :

362	715	826	987	567
571	157	628	789	657
263	751	682	879	765
623	286	307	978	576
175	268	703	798	675
517	862	370	897	756

Write in figures the following numbers :

One hundred twenty-nine.
Two hundred thirty-six.
Two hundred twenty-four.
Two hundred twenty-two.
Five hundred nineteen.
Seven hundred thirteen.
Six hundred eighteen.
Nine hundred eleven.
Three hundred twelve.
Three hundred sixteen.

One hundred nine.
Seven hundred eight.
Five hundred six.
Four hundred seven.
Three hundred five.
Two hundred four.
Four hundred three.
Three hundred two.
Four hundred one.
Four hundred ten.

In any number containing hundreds, tens, and ones,

The ones are called **units of the first order.**

The tens are called **units of the** second **order.**

The hundreds are called **units of the third order.**

Remember that **any** standard by which we count or measure is called a **unit.**

Find the sums:

128	136	215	320	357
362	204	327	267	198
416	473	296	376	276

317	218	375	427	576
207	219	293	291	197
327	397	189	198	189

229	379	263	327	183
292	125	362	279	136
376	268	185	202	181

Find the remainders:

362	416	473	327	355
128	137	279	158	278

811	821	725	527	283
624	583	258	279	196

615	913	916	874	767
209	467	529	389	488

531	451	937	873	726
253	184	690	565	339

657	765	675	897	703
567	576	386	798	370

862	517	726	904	703
218	175	528	208	307

FORTY-TWO.   42.

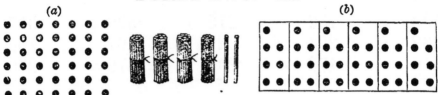

How many dots in each row of dots marked (a)?
How many rows?
How many dots in the six rows?
How many dots, then, are 6 times 7 dots?
How many dots in each column of dots?
How many columns?
How many dots in the seven columns?
How many dots, then, are 7 times 6 dots?
How many 7's in 42?   How many 6's in 42?

$6 \times 7 = ?$ $7 \times 6 = ?$ $42 \div 7 = ?$ $42 \div 6 = ?$

At 6 cents a pound, what will 7 pounds of sugar cost? 6 pounds? 5 pounds? 4 pounds?

At 7 cents a quart, how many quarts of blueberries can you buy for 42 cents?

At 6 dollars a ton, how many tons of coal can be bought for 42 dollars?

At 7 cents each, what will 6 melons cost?
Count by 2's to 42.   Count by 3's to 42.
Count by 4's to 40.   Count by 5's to 40.
Count by 6's to 42.   Count by 7's to 42.
How many 7's in 28? 35? 42? 21? 14?
How many 6's in 24? 30? 36? 42? 18?
How many 5's in 25? 30? 35? 40? 20?

FORTY-FIVE. 45.

How many dots in each row of dots marked (a)?

How many rows?

How many dots in the five rows?

How many dots, then, are 5 times 9 dots?

How many dots in each column of dots?

How many columns?

How many dots in the nine columns?

How many dots, then, are 9 times 5 dots?

How many 9's in 45?   How many 5's in 45?

$5 \times 9 = ?$ $9 \times 5 = ?$ $45 \div 5 = ?$ $45 \div 9 = ?$

At 5 cents a pound, how many pounds of sugar can be bought for 45 cents? for 40 cents?

At 9 cents a pound, how many pounds of candy can be bought for 45 cents? for 36 cents?

Copy, and write the answers:

$2 \times 2 = ?$	$3 \times 2 = ?$	$4 \times 2 = ?$	$5 \times 2 = ?$
$2 \times 3 = ?$	$3 \times 3 = ?$	$4 \times 3 = ?$	$5 \times 3 = ?$
$2 \times 4 = ?$	$3 \times 4 = ?$	$4 \times 4 = ?$	$5 \times 4 = ?$
$2 \times 5 = ?$	$3 \times 5 = ?$	$4 \times 5 = ?$	$5 \times 5 = ?$
$2 \times 6 = ?$	$3 \times 6 = ?$	$4 \times 6 = ?$	$5 \times 6 = ?$
$2 \times 7 = ?$	$3 \times 7 = ?$	$4 \times 7 = ?$	$5 \times 7 = ?$
$2 \times 8 = ?$	$3 \times 8 = ?$	$4 \times 8 = ?$	$5 \times 8 = ?$
$2 \times 9 = ?$	$3 \times 9 = ?$	$4 \times 9 = ?$	$5 \times 9 = ?$

Copy, and write the answers:

$4 \div 2 = ?$	$6 \div 3 = ?$	$8 \div 4 = ?$	$10 \div 5 = ?$
$6 \div 2 = ?$	$9 \div 3 = ?$	$12 \div 4 = ?$	$15 \div 5 = ?$
$8 \div 2 = ?$	$12 \div 3 = ?$	$16 \div 4 = ?$	$20 \div 5 = ?$
$10 \div 2 = ?$	$15 \div 3 = ?$	$20 \div 4 = ?$	$25 \div 5 = ?$
$12 \div 2 = ?$	$18 \div 3 = ?$	$24 \div 4 = ?$	$30 \div 5 = ?$
$14 \div 2 = ?$	$21 \div 3 = ?$	$28 \div 4 = ?$	$35 \div 5 = ?$
$16 \div 2 = ?$	$24 \div 3 = ?$	$32 \div 4 = ?$	$40 \div 5 = ?$
$18 \div 2 = ?$	$27 \div 3 = ?$	$36 \div 4 = ?$	$45 \div 5 = ?$
$20 \div 2 = ?$	$30 \div 3 = ?$	$40 \div 4 = ?$	$50 \div 5 = ?$

Find

$\frac{1}{2}$ of 4.	$\frac{1}{3}$ of 6.	$\frac{1}{4}$ of 8.	$\frac{1}{5}$ of 10.	$\frac{1}{6}$ of 12.
$\frac{1}{2}$ of 6.	$\frac{1}{3}$ of 9.	$\frac{1}{4}$ of 12.	$\frac{1}{5}$ of 15.	$\frac{1}{6}$ of 18.
$\frac{1}{2}$ of 8.	$\frac{1}{3}$ of 12.	$\frac{1}{4}$ of 16.	$\frac{1}{5}$ of 20.	$\frac{1}{6}$ of 24.
$\frac{1}{2}$ of 10.	$\frac{1}{3}$ of 15.	$\frac{1}{4}$ of 20.	$\frac{1}{5}$ of 25.	$\frac{1}{6}$ of 30.
$\frac{1}{2}$ of 12.	$\frac{1}{3}$ of 18.	$\frac{1}{4}$ of 24.	$\frac{1}{5}$ of 30.	$\frac{1}{6}$ of 36.
$\frac{1}{2}$ of 14.	$\frac{1}{3}$ of 21.	$\frac{1}{4}$ of 28.	$\frac{1}{5}$ of 35.	$\frac{1}{6}$ of 42.
$\frac{1}{2}$ of 16.	$\frac{1}{3}$ of 24.	$\frac{1}{4}$ of 32.	$\frac{1}{5}$ of 40.	$\frac{1}{7}$ of 42.
$\frac{1}{2}$ of 18.	$\frac{1}{3}$ of 27.	$\frac{1}{4}$ of 36.	$\frac{1}{5}$ of 45.	$\frac{1}{7}$ of 35.
$\frac{1}{2}$ of 20.	$\frac{1}{3}$ of 30.	$\frac{1}{4}$ of 40.	$\frac{1}{5}$ of 50.	$\frac{1}{7}$ of 28.

When we multiply one number by another, the result is called the **product**; the number multiplied is called the **multiplicand**; and the number by which we multiply is called the **multiplier.**

Name two numbers whose product is: **15**; **12**; **18**; **24**; **21**; **32**; **28**; **25**; **35**; **45**; **42**; **27**; **20.**

The product of two equal numbers is called a **square number.** With 16 buttons make a square.

Multiply 234 by 2.

Write the multiplicand   234
Under the *ones* write the multiplier   2
Draw a line below.   ———
Multiply in order the *ones*, *tens*, and *hundreds*, and write   468
the result at each step : Twice 4 ones are 8 ones, twice 3 tens are 6
tens, twice 2 hundreds are 4 hundreds. The product, therefore, is 468.

Find the products :

342	123	243	334	321	424
2	2	2	2	2	2

123	132	323	213	312	212
3	3	3	3	3	3

111	112	121	211	212	222
4	4	4	4	4	4

When we divide one number by another, the
result is called the **quotient**; the number divided
is called the **dividend**; and the number by which
we divide is called the **divisor**.

Divide 648 by 2.

Write the **divisor** at the left of the **dividend** with a   2)648
curved line between them, and draw a line underneath.   ———
Divide in order the *hundreds*, *tens*, and *ones*, and write   324
the result at each step : 2 in 6 hundreds, 3 hundreds ; 2 in 4 tens,
2 tens ; 2 in 8 ones, 4 ones. The quotient, therefore, is 324.

Find the quotients :

2)428	2)684	2)468	2)864	2)248
3)369	3)639	3)396	3)693	3)936
3)963	4)444	4)484	4)448	4)844

ROMAN NUMERALS.

The Roman method of writing numbers uses these seven capital letters:

I = 1;   V = 5;   X = 10;   L = 50;
C = 100;   D = 500;   M = 1000.

Other numbers are written by putting two or more of these letters together.

A letter written **before another of** greater value signifies the **difference** of the values of the letters used.

Thus, IV = 4;   IX = 9;   XL = 40;   XC = 90.

A letter written after another **of the same or** greater value signifies the **sum** of the values of the letters used. Thus,

VI = 6;   XI = 11;   LX = 60;   CX = 110;
II = 2;   III = 3;   VII = 7;   VIII = 8;
XX = 20;   XXX = 30;   LXX = 70;   CCC = 300·

Numbers from 10 to 20 are written :

11 = X + I   = XI;   16 = X + VI   = XVI;
12 = X + II = XII;   17 = X + VII = XVII;
13 = X + III = XIII;   18 = X + VIII = XVIII;
14 = X + IV = XIV;   19 = X + IX   = XIX.
15 = X + V = XV;

In like manner:

25 = ·XX + V = XXV;   46 = XL + VI = XLVI;
29 = XX + IX = XXIX;   69 = LX + IX = LXIX.

Complete with Roman numerals:

1 =	11 =	21 =	35 =	64 =
2 =	12 =	22 =	45 =	36 =
3 =	13 =	23 =	55.=	46 =
4 =	14 =	24 =	65 =	77 =
5 =	15 =	25 =	75 =	88 =
6 =	16 =	26 =	85 =	97 =
7 =	17 =	27 =	95 =	39 =
8 =	18 =	28 =	34 =	98 =
9 =	19 =	29 =	44 =	89 =
10 =	20 =	30 =	54 =	99 =

Complete with figures:

I =	XV =	XXIX =	XCI =
II =	XVI =	XXX =	XCII =
III =	XVII =	XL =	XCIII =
IV =	XVIII =	XLV =	XCIV =
V =	XIX =	LI =	XCV =
VI =	XX =	LII =	XCVI =
VII =	XXI =	LIII =	XCVII =
VIII =	XXII =	LIV =	XCVIII =
IX =	XXIII =	LV =	XCIX =
X =	XXIV =	LVI =	CVIII =
XI =	XXV =	LVII =	CL =
XII =	XXVI =	LVIII =	CCIX =
XIII =	XXVII =	LIX =	CCXX =
XIV =	XXVIII =	LX =	CCXLV =

MEASURE OF TIME.

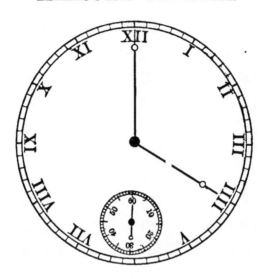

When the smallest hand of the clock has gone round the little circle, a *minute* has passed.

The little circle has 60 spaces, and the hand goes over one space every *second*. Hence,

Sixty seconds make a minute.

When the longest hand of the clock has gone round the large circle, an *hour* has passed.

How many spaces are marked on the large circle?

The longest hand goes over one space every *minute*. Hence,

Sixty minutes make an hour.

The letters I, II, etc., mark the hour spaces.

How many hours have passed when the hour-hand has gone entirely round the face of the clock?

The hour-hand goes round twice from sunrise to sunrise. Hence,

Twenty-four hours make a day.

How many minutes in a half of an hour? in a quarter of an hour? in a third of an hour? in three-quarters of an hour?

What part of an hour are 30 minutes? 15 minutes? 20 minutes? 10 minutes? 45 minutes?

How many hours in a half of a day? in a quarter of a day? in a third of a day?

What time of day is shown on the clock-face?

What time of day will be shown on the clock-face when the minute-hand reaches I? II? III? IIII? V? VI? VII? VIII? IX? X? XI? XII?

What time of day will be shown on the clock-face when the minute-hand is one minute-space beyond I? II? III? V? VI? VIII? IX? X? XI?

What time of day will be shown on the clock-face when the minute-hand is two minute-spaces beyond I? II? III? IIII? VI? IX? X? XI?

What time of day will be shown on the clock-face when the minute-hand is three minute-spaces beyond I? III? V? VII? IX? X? XI?

What time of day will be shown on the clock-face when the minute-hand is four minute-spaces beyond II? III? V? VI? VII? VIII? IX? X?

What time of day will be shown on the clock-face when the minute-hand is at XII and the hour-hand at I? II? III? V? VI? VII? VIII? IX?

At what letters does the minute-hand point at half-past four? at quarter-past four? at quarter of five? at 20 minutes to five?

If a man works 8 hours a day, what part of the day (24 hours) does he work?

What part of 24 hours are 4 hours? 6 hours? 8 hours? 12 hours? 2 hours?

If a man can dig one-quarter of a certain ditch in 8 hours, how many hours will it take him to dig the whole ditch?

If 2 men can mow a certain field in 8 days, how many days will it take one man to mow it?

If one man can mow a certain field in 24 days, how many men will it take to mow the field in 6 days? in 4 days? in 8 days? in 3 days?

How many minutes are there in 2 hours? in 3 hours? in 4 hours? in 5 hours? in 6 hours?

How many seconds are there in 2 minutes? in 4 minutes? in 5 minutes? in 6 minutes?

What part of a minute are 30 seconds? 15 seconds? 12 seconds? 20 seconds? 40 seconds? 45 seconds? 50 seconds?

If a man walks a mile in 20 minutes, how many miles at that rate will he walk in an hour?

If a man walks a mile in 15 minutes, how many miles at that rate will he walk in an hour?

At the rate of one mile in 10 minutes, how many miles will a horse go in an hour?

At the rate of one mile in 6 minutes, how many miles will a horse go in one hour?

At the rate of one mile in 2 minutes, how many miles will a railway train go in an hour?

Part III.

LESSON 1.

FORTY-EIGHT. 48.

How many dots are there in each row?

How many rows are there?

How many dots in the six rows?

How many dots, then, are 6 times 8 dots?

How many dots are there in each column?

How many columns are there?

How many dots in the eight columns?

How many dots, then, are 8 times 6 dots?

$6 \times 8 = ?$	$8 \times 6 = ?$	$48 \div 8 = ?$	$48 \div 6 = ?$
$\frac{1}{6}$ of $48 = ?$	$\frac{1}{8}$ of $48 = ?$	$\frac{1}{4}$ of $48 = ?$	$\frac{1}{3}$ of $48 = ?$

At 6 dollars a ton, what will 8 tons of coal cost?

At 8 dollars apiece, what will 6 hats cost?

If a cow gives 8 quarts of milk a day, in how many days will she give 48 quarts?

111

FORTY-NINE. 49.

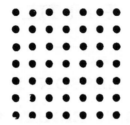

How many dots are there in each row?

How many rows are there?

How many dots in the seven rows?

How many dots, then, are 7 times 7 dots?

Count by 7's to 49.   Count by 8's to 48.

$7 \times 7 = ?$	$49 \div 7 = ?$	$\frac{1}{7}$ of $49 = ?$	$2 \times 7 = ?$
$3 \times 7 = ?$	$4 \times 7 = ?$	$5 \times 7 = ?$	$6 \times 7 = ?$
$7 + 7 = ?$	$49 - 7 = ?$	$42 - 7 = ?$	$35 - 7 = ?$
$28 - 7 = ?$	$21 - 7 = ?$	$14 - 7 = ?$	$7 - 7 = ?$

At 7 cents a pound, what will 7 pounds of rice cost? 6 pounds? 5 pounds? 4 pounds? 3 pounds?

Copy and subtract:

418	219	607	729	839
−166	−184	−235	−327	−655
905	806	704	603	502
−461	−235	−194	−152	−171
213	314	415	516	617
−151	−182	−193	−264	−255
526	425	324	635	639
−275	−283	−193	−383	−379

FIFTY-FOUR. 54.

How many dots are there in each row?

How many rows are there?

How many dots in the six rows?

How many dots, then, are 6 times 9 dots?

How many dots are there in each column?

How many columns are there?

How many dots in the nine columns?

How many dots, then, are 9 times 6 dots?

Count by 6's to 54.   Count by 9's to 54.

$6 \times 9 = ?$	$9 \times 6 = ?$	$54 \div 6 = ?$	$54 \div 9 = ?$
$\frac{1}{2}$ of $54 = ?$	$\frac{1}{3}$ of $54 = ?$	$\frac{1}{6}$ of $54 = ?$	$\frac{1}{9}$ of $54 = ?$
$2 \times 6 = ?$	$6 \times 6 = ?$	$12 \div 6 = ?$	$36 \div 6 = ?$
$3 \times 6 = ?$	$7 \times 6 = ?$	$18 \div 6 = ?$	$42 \div 6 = ?$
$4 \times 6 = ?$	$8 \times 6 = ?$	$24 \div 6 = ?$	$48 \div 6 = ?$
$5 \times 6 = ?$	$9 \times 6 = ?$	$30 \div 6 = ?$	$54 \div 6 = ?$

How many tens and how many ones in 54?

$54 - 6 = ?$	$48 - 6 = ?$	$42 - 6 = ?$	$36 - 6 = ?$
$30 - 6 = ?$	$24 - 6 = ?$	$18 - 6 = ?$	$12 - 6 = ?$
$54 - 9 = ?$	$45 - 9 = ?$	$36 - 9 = ?$	$27 - 9 = ?$

At 6 cents a quart, what will 9 quarts of milk cost? 8 quarts? 7 quarts? 6 quarts? 4 quarts?

At 9 cents a pint, what will 6 pints of sirup cost? 5 pints? 4 pints? 3 pints? 2 pints?

If we divide 25 by 4, we have 6 for the quotient and 1 for the remainder.

The quotient and remainder may be written as a complete quotient, thus, 6¼.

In this quotient, the part ¼ is written by writing the remainder above the divisor with a line between them.

Divide, and write the complete quotient under the dividend in each case :

2) 13	3) 20	3) 29	5) 21	6) 13
2) 15	3) 22	4) 21	5) 27	5) 32
2) 17	3) 23	4) 23	5) 33	6) 39
2) 19	3) 25	4) 33	5) 34	6) 40
3) 19	3) 26	4) 35	5) 37	6) 47
3) 17	3) 28	4) 37	5) 44	6) 53

2) 123	3) 123	4) 124	6) 126
2) 143	3) 153	4) 128	6) 128
2) 167	3) 157	4) 160	6) 186
2) 165	3) 159	4) 166	6) 180
2) 169	3) 127	4) 168	6) 248
2) 182	3) 128	4) 204	6) 249
2) 184	3) 187	4) 247	6) 306
2) 187	3) 189	4) 289	6) 368

FIFTY-SIX. 56.

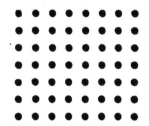

How many dots are there in each row ?

How many rows are there ?

How many dots in the seven rows ?

How many dots, then, are 7 times 8 dots ?

How many dots in each column ?

How many columns are there ?

How many dots in the éight columns ?

How many dots, then, are 8 times 7 dots ?

$7 \times 8 = ?$ $8 \times 7 = ?$ $56 \div 7 = ?$ $56 \div 8 = ?$

$\frac{1}{7}$ of $56 = ?$ $\frac{1}{8}$ of $56 = ?$ $\frac{1}{4}$ of $56 = ?$ $\frac{1}{2}$ of $56 = ?$

If $\frac{1}{4}$ of 56 is 14, and $\frac{1}{8}$ of 56 is 7, how many *eighths* of 56 are equal to $\frac{1}{4}$ of 56 ?

How many *eighths* of 56 are equal to $\frac{2}{4}$ of 56 ?

Count by 8's to 56.   Count by 7's to 56.

If a man works 8 hours a day, how many hours will he work in 5 days ? in 6 days ? in 7 days ?

What will 7 yards of print cost, at 8 cents a yard ? at 7 cents a yard ? at 6 cents a yard ?

At 8 cents a yard, how many yards of cambric can be bought for 40 cents ? for 48 cents ?

At 7 dollars a ton, how many tons of coal can be bought for 49 dollars ? for 56 dollars ?

SIXTY-THREE. 63.

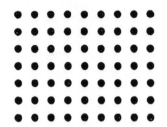

How many dots are there in each row?

How many rows are there?

How many dots in the seven rows?

How many dots, then, are 7 times 9 dots?

How many dots are there in each column?

How many columns are there?

How many dots in the nine columns?

How many dots, then, are 9 times 7 dots?

$7 \times 9 = ?$ $9 \times 7 = ?$ $63 \div 7 = ?$ $63 \div 9 = ?$

$\frac{1}{7}$ of $63 = ?$ $\frac{1}{9}$ of $63 = ?$ $\frac{1}{3}$ of $63 = ?$ $\frac{2}{3}$ of $63 = ?$

How many *ninths* of 63 are equal to $\frac{1}{3}$ of 63?

Count by 7's to 63.   Count by 9's to 63.

At 9 cents a foot, what will 7 feet of lead pipe cost? 6 feet? 4 feet? 5 feet? 3 feet?

How many days are there in 9 weeks?

At 7 dollars a week, how many weeks' board can be had for 56 dollars? for 63 dollars?

At 9 cents a quart, how many quarts of cranberries can be bought for 54 cents? for 63 cents?

How many quarts of oats in 7 peeks of oats?

How many dozen eggs in 48 eggs?

How many gallons of milk in 36 quarts of milk?

$7 \times 2 = ?$	$7 \times 6 = ?$	$14 \div 7 = ?$	$42 \div 7 = ?$
$7 \times 3 = ?$	$7 \times 7 = ?$	$21 \div 7 = ?$	$49 \div 7 = ?$
$7 \times 4 = ?$	$7 \times 8 = ?$	$28 \div 7 = ?$	$56 \div 7 = ?$
$7 \times 5 = ?$	$7 \times 9 = ?$	$35 \div 7 = ?$	$63 \div 7 = ?$

Copy, and find the products:

12	12	11	11	11	11	11
3	4	5	6	7	8	9

41	41	41	41	41	41	41
3	4	5	6	7	8	9

60	60	60	60	60	60	60
3	4	5	6	7	8	9

31	31	31	31	31	31	31
3	4	5	6	7	8	9

70	70	70	70	70	70	70
3	4	5	6	7	8	9

80	80	80	80	80	50	60
3	4	5	6	7	8	9

91	91	91	91	91	71	61
3	4	5	6	7	8	9

80	80	80	80	80	80	80
3	4	5	6	7	8	9

81	71	61	91	51	41	31
7	8	9	6	8	7	6

A fly has 6 legs. How many legs have 9 flies?

A spider has 8 legs. How many legs have 7 spiders? 6 spiders? 4 spiders? 3 spiders?

An ox has 8 hoofs. How many hoofs have 6 oxen? 5 oxen? 4 oxen? 3 oxen?

A man bought 9 cords of wood at 4 dollars a cord, and gave 4 ten-dollar bills in payment. How much change should he receive?

James had 7 cents, and his father gave him six times as much. How many cents had he then?

Ernest has 9 five-cent pieces and 3 cents. How much money has he?

What will 9 sheep cost, at 6 dollars each?

At 7 cents a yard, what will 9 yards of cotton cloth cost? What will 8 yards cost?

A farmer sold 9 lambs for 45 dollars. How much apiece did he get for them?

How many lengths of 9 yards each can be cut from a piece of silk 63 yards long?

In a schoolroom there were 63 seats arranged in 7 rows. How many seats in each row?

Find the cost of a dozen peaches at 3 for 5 cents.

Find the cost of a dozen pears at 3 for 4 cents.

A bushel of oats weighs 32 pounds. How many pounds will a peek weigh? 3 peeks?

A bushel of corn weighs 56 pounds. How many pounds will a peek weigh? 2 peeks?

At 56 cents a peek, how much must be paid for a quart of beans? 2 quarts? 4 quarts? 6 quarts?

SIXTY-FOUR. **64.**

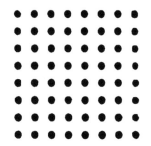

How many dots are there in each row ?
How many rows are there ?
How many dots in the eight rows ?
How many dots, then, are 8 times 8 dots ?
Count by 8's to 64.

$8 \times 8 = ?$　　　　$64 \div 8 = ?$　　　　$\frac{1}{8}$ of 64 = ?

A man receives 8 dollars a week for work. How much does he receive in 8 weeks ?

There are 8 pints in a gallon. How many pints are there in 8 gallons ? in 7 gallons ?

When flour is 6 dollars a barrel, what will 8 barrels cost ? 9 barrels ? 7 barrels ? 4 barrels ?

When blueberries are 8 cents a quart, what will 7 quarts cost ? 8 quarts ? 6 quarts ? 5 quarts ?

At 7 cents a quart, what will a peek of beans cost ? What will 9 quarts cost ? What will 6 quarts cost ? What will 4 quarts cost ?

If a freight train averages 8 miles an hour, in how many hours will it run 64 miles ? 56 miles ?

$64 - 8 = ?$	$56 - 8 = ?$	$48 - 8 = ?$	$40 - 8 = ?$
$32 - 8 = ?$	$24 - 8 = ?$	$16 - 8 = ?$	$8 - 8 = ?$

SEVENTY-TWO.   72.

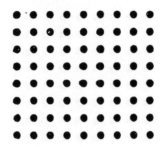

How many dots are there in each row?

How many rows are there?

How many dots in the eight rows?

How many dots, then, are 8 times 9 dots?

How many dots are there in each column?

How many columns are there?

How many dots in the nine columns?

How many dots, then, are 9 times 8 dots?

$8 \times 9 = ?$ $9 \times 8 = ?$ $72 \div 8 = ?$ $72 \div 9 = ?$

$\frac{1}{8}$ of $72 = ?$ $\frac{1}{9}$ of $72 = ?$ $\frac{1}{6}$ of $72 = ?$ $\frac{1}{4}$ of $72 = ?$

At 8 dollars apiece, what will be the cost of 9 calves? 7 calves? 8 calves? 6 calves?

At 9 cents a yard, what will be the cost of 8 yards of cambric? 7 yards? 6 yards?

A farmer sold 8 calves for 72 dollars. How much did he get apiece?

If 9 yards of muslin cost 72 cents, what is the price of the muslin a yard?

How many 9's in 36? in 54? in 63? in 45? in 72? in 27? in 18?

How many dozen in 24? in 36? in 48? in 72?

$2 \times 8 = ?$	$6 \times 8 = ?$	$16 \div 2 = ?$	$48 \div 6 = ?$
$3 \times 8 = ?$	$7 \times 8 = ?$	$24 \div 3 = ?$	$56 \div 7 = ?$
$4 \times 8 = ?$	$8 \times 8 = ?$	$32 \div 4 = ?$	$64 \div 8 = ?$
$5 \times 8 = ?$	$9 \times 8 = ?$	$40 \div 5 = ?$	$72 \div 9 = ?$

$\frac{1}{8}$ of $16 = ?$	$\frac{1}{8}$ of $32 = ?$	$\frac{1}{8}$ of $48 = ?$	$\frac{1}{8}$ of $64 = ?$
$\frac{1}{2}$ of $16 = ?$	$\frac{1}{4}$ of $32 = ?$	$\frac{1}{6}$ of $48 = ?$	$\frac{1}{4}$ of $64 = ?$
$\frac{1}{8}$ of $24 = ?$	$\frac{1}{8}$ of $40 = ?$	$\frac{1}{8}$ of $56 = ?$	$\frac{1}{8}$ of $72 = ?$
$\frac{1}{3}$ of $24 = ?$	$\frac{1}{5}$ of $40 = ?$	$\frac{1}{7}$ of $56 = ?$	$\frac{1}{9}$ of $72 = ?$

Add :

23	31	27	36	47	75
35	49	33	67	51	24
47	36	29	73	68	37
72	53	32	21	33	22

36	67	76	89	98	57
84	74	88	37	65	84
39	38	31	53	29	37
46	21	19	27	37	18

Find the differences :

225	313	321	337	235
87	56	28	89	88

312	482	563	671	817
147	279	392	289	465

476	567	675	576	637
279	378	387	378	239

EIGHTY-ONE. 81.

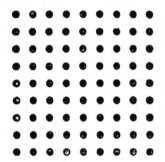

How many dots are there in each row?

How many rows are there?

How many dots in the nine rows?

How many dots, then, are 9 times 9 dots?

Count by 9's to 81. $9 \times 9 = ?$ $81 \div 9 = ?$

If it takes 9 yards of cloth for a dress, how many yards will be required for 9 dresses?

If a family uses 9 pounds of sugar a week, how many weeks will 81 pounds last the family?

If it takes 7 eggs for a cake, how many eggs will be required for 9 cakes?

How many days are there in 9 weeks?

If it takes 9 yards of print for a dress, how many dresses can be made from 54 yards?

If you sleep 8 hours every night, how many hours will you sleep in 9 nights?

$2 \times 9 = ?$	$6 \times 9 = ?$	$18 \div 9 = ?$	$54 \div 9 = ?$
$3 \times 9 = ?$	$7 \times 9 = ?$	$27 \div 9 = ?$	$63 \div 9 = ?$
$4 \times 9 = ?$	$8 \times 9 = ?$	$36 \div 9 = ?$	$72 \div 9 = ?$
$5 \times 9 = ?$	$9 \times 9 = ?$	$45 \div 9 = ?$	$81 \div 9 = ?$

MULTIPLICATION TABLE.

2 TIMES	3 TIMES	4 TIMES	5 TIMES
1 ARE 2	1 ARE 3	1 ARE 4	1 ARE 5
2 ARE 4	2 ARE 6	2 ARE 8	2 ARE 10
3 ARE 6	3 ARE 9	3 ARE 12	3 ARE 15
4 ARE 8	4 ARE 12	4 ARE 16	4 ARE 20
5 ARE 10	5 ARE 15	5 ARE 20	5 ARE 25
6 ARE 12	6 ARE 18	6 ARE 24	6 ARE 30
7 ARE 14	7 ARE 21	7 ARE 28	7 ARE 35
8 ARE 16	8 ARE 24	8 ARE 32	8 ARE 40
9 ARE 18	9 ARE 27	9 ARE 36	9 ARE 45

6 TIMES	7 TIMES	8 TIMES	9 TIMES
1 ARE 6	1 ARE 7	1 ARE 8	1 ARE 9
2 ARE 12	2 ARE 14	2 ARE 16	2 ARE 18
3 ARE 18	3 ARE 21	3 ARE 24	3 ARE 27
4 ARE 24	4 ARE 28	4 ARE 32	4 ARE 36
5 ARE 30	5 ARE 35	5 ARE 40	5 ARE 45
6 ARE 36	6 ARE 42	6 ARE 48	6 ARE 54
7 ARE 42	7 ARE 49	7 ARE 56	7 ARE 63
8 ARE 48	8 ARE 56	8 ARE 64	8 ARE 72
9 ARE 54	9 ARE 63	9 ARE 72	9 ARE 81

Robert bought 2 postage stamps at 3 cents apiece. How much did he pay for them?

A buggy has 4 wheels. How many wheels are needed for 2 buggies?

At 5 cents each, how much will 2 car tickets cost?

At 6 cents a quart, how much will 2 quarts of peanuts cost?

At 7 cents a pound, what will be the cost of 2 pounds of loaf sugar?

At 8 cents a yard, what will 2 yards of calico cost?

At 9 cents a cake, what will 2 cakes of soap cost?

If a horse goes 9 miles an hour for 3 hours, how many miles will he go in all?

A box has eight corners. How many corners have 3 boxes together?

If a pair of boots costs 7 dollars, how many dollars will 3 pairs of boots cost?

If an orange costs 3 cents, how many oranges can you buy for 21 cents? for 27 cents? for 24 cents? for 18 cents? for 12 cents?

At 3 cents apiece, how much will 4 oranges cost?

If a hat costs 4 dollars, how much will 4 hats cost?

At 5 dollars a barrel, what will be the cost of 4 barrels of flour?

At 6 cents a quart, what will be the cost of a gallon of milk?

At 7 cents apiece, what will be the cost of 4 yard-sticks?

At 9 dollars a barrel, what will be the cost of 4 barrels of brown sugar?

At 8 dollars a load, what will 4 loads of bricks cost?

A farmer sold 5 pigs for 3 dollars apiece. How much did he get for his pigs?

A farmer sold 6 barrels of apples at 3 dollars a barrel. How much did the 6 barrels bring?

If one desk has 8 drawers, how many drawers will 5 desks of the same pattern have?

How many yards long is a piece of cloth that is 24 feet long?

How many pints of milk will a two-gallon can hold?

How many quarts of oysters in 6 gallons of oysters?

The cook used 2 dozen eggs in making six puddings. How many eggs on the average did she use for each pudding?

At 5 cents apiece, how many bananas can be bought for 30 cents?

At 4 cents apiece, how many oranges can be bought for 24 cents? At 3 cents apiece, how many can be bought for 24 cents?

At 6 cents a quart, how many quarts of berries can you buy for 18 cents? for 36 cents? for 30 cents? for 24 cents? for 42 cents? for 48 cents?

If a quarter of a pound of candy costs 9 cents, what will a pound cost?

I have 40 cents in 5-cent pieces. How many 5-cent pieces have I?

At 10 cents a quire, how many quires of paper can be bought for 40 cents?

James has 50 cents in 10-cent pieces. How many 10-cent pieces has he?

If 36 pounds of starch are put up in 4-pound packages, how many packages will there be?

John has 54 cents. How many quarts of peanuts can he buy at 6 cents a quart?

Emma has 54 cents. How many yards of ribbon can she buy at 9 cents a yard?

At 8 cents a quart, what will 7 quarts of berries cost?

At 7 cents a yard, what will 8 yards of cloth cost?

Robert has 56 cents. How many packages of candy can he buy if each package is 8 cents?

An orchard has 56 trees, and there are 7 equal rows. How many trees in each row?

A certain schoolroom has 7 rows of desks, with 9 desks in each row. How many desks in the 7 rows?

A man can build 9 yards of fence in a day. How many days will it take him to build 63 yards?

A dealer sold 7 plows for 63 dollars. What was the price of one plow?

A man took 6 eggs at a time seven times from a box of eggs. How many eggs did he take out?

If a ton of coal costs 6 dollars, how much will 9 tons cost?

If a cord of oak wood is worth 7 dollars, how much will 9 cords cost?

On a table there are 9 plates, and each plate has 9 peaches. How many peaches are on the table?

If one dozen buttons cost 8 cents, how much will 9 dozen buttons cost?

Ernest has 64 buttons. How many rows of 8 buttons each can he make?

If a box of butter weighs 7 pounds, how much will 8 boxes weigh?

It takes 6 candles to weigh a pound. How many pounds will 54 candles weigh?

At 7 dollars a pair, how many pairs of boots can be bought for 56 dollars? for 63 dollars?

If three men together earn 9 dollars a day, in how many days will they earn 54 dollars?

If a man earns 8 dollars a week, in how many weeks will he earn 56 dollars?

There are 6 working days in a week. How many working days are there in 7 weeks?

If 9 persons ride in a coach, how many coaches will be required to carry 72 persons?

If a man has 48 horses, how many 6-horse teams can he form?

How many pecks in 56 quarts?

Copy and multiply:

94	43	62	51	71	81
2	3	4	5	7	8

71	91	81	61	31	92
9	8	7	6	5	4

920	930	610	710	910	810
3	2	9	8	7	6

210	310	710	910	810	920
9	8	7	6	5	4

622	911	711	911	811	911
4	5	6	7	8	9

913	944	811	810	101	901
3	2	5	7	8	9

Copy and divide:

2) 266	3) 273	4) 364	5) 455	6) 546
7) 567	8) 648	9) 729	7) 637	5) 405
3) 213	4) 484	2) 468	6) 606	8) 808
5) 550	6) 546	9) 909	8) 568	7) 777
7) 567	9) 549	4) 884	5) 500	8) 568
6) 546	7) 721	8) 856	9) 972	4) 836

Count to a number greater than 100:

By 2's, beginning with 1; with 2.

By 3's, beginning with 1; with 2; with 3.

By 4's, beginning with 1; with 2; with 3; with 4.

By 5's, beginning with 1; with 2; with 3; with 4; with 5.

By 6's, beginning with 1; with 2; with 3; with 4; with 5; with 6.

By 7's, beginning with 1; with 2; with 3; with 4; with 5; with 6; with 7.

By 8's, beginning with 1; with 2; with 3; with 4; with 5; with 6; with 7; with 8.

By 9's, beginning with 1; with 2; with 3; with 4; with 5; with 6; with 7; with 8; with 9.

NOTE. Practice the above drill-exercise until every pupil can go through it readily.

In counting by 2's, beginning with 2, we obtain 2, 4, 6, 8, 10, 12, 14, 16, 18, 20, etc.

These numbers are called **even numbers**.

In counting by 2's, beginning with 1, we obtain 1, 3, 5, 7, 9, 11, 13, 15, 17, 19, etc.

These numbers are called **odd numbers**.

With what figures do even numbers end?

With what figures do odd numbers end?

Does any even number when divided by 2 give a remainder?

Which of the following numbers are odd, and which even?

5, 7, 10, 25, 36, 38, 47, 50, 51, 55.

How many 11's in 22? in 33? in 44? in 55? in 66? in 77? in 88? in 99? in 110? in 121? in 132?

How many 12's in 24? in 36? in 48? in 60? in 72? in 84? in 96? in 108? in 120? in 132? in 144?

How many eggs are 2 dozen eggs? 3 dozen? 4 dozen? 5 dozen? 6 dozen? 7 dozen? 8 dozen? 9 dozen? 10 dozen? 11 dozen? 12 dozen?

$2 \times 11 =$?	$7 \times 11 =$?	$2 \times 12 =$?	$7 \times 12 =$?
$11 \times 2 =$?	$11 \times 7 =$?	$12 \times 2 =$?	$12 \times 7 =$?
$3 \times 11 =$?	$8 \times 11 =$?	$3 \times 12 =$?	$8 \times 12 =$?
$11 \times 3 =$?	$11 \times 8 =$?	$12 \times 3 =$?	$12 \times 8 =$?
$4 \times 11 =$?	$9 \times 11 =$?	$4 \times 12 =$?	$9 \times 12 =$?
$11 \times 4 =$?	$11 \times 9 =$?	$12 \times 4 =$?	$12 \times 9 =$?
$5 \times 11 =$?	$10 \times 11 =$?	$5 \times 12 =$?	$10 \times 12 =$?
$11 \times 5 =$?	$11 \times 10 =$?	$12 \times 5 =$?	$11 \times 12 =$?
$6 \times 11 =$?	$11 \times 11 =$?	$6 \times 12 =$?	$12 \times 11 =$?
$11 \times 6 =$?	$11 \times 12 =$?	$12 \times 6 =$?	$12 \times 12 =$?

At 12 cents each, what is the cost of 11 slates?

At 12 dollars each, what is the cost of 12 coats?

If a man works 9 hours a day, how many hours will he work in 2 weeks? in one week and a half?

Twelve months make a year.

How many months in 2 years? in 7 years?

How many years in 36 months? in 96 months?

Thirty-six inches make a yard.

How many inches in 1 yard? in $\frac{5}{6}$ of a yard? in $\frac{3}{4}$ of a yard? in $\frac{2}{3}$ of a yard? in $\frac{1}{3}$ of a yard and $\frac{1}{4}$ of a foot? in $\frac{1}{2}$ of a yard and $\frac{1}{2}$ of a foot?

1 SQUARE FOOT.

This square represents a square foot.

How many inches long is a side of the square?

How many square inches are there in the square?

144 square inches make 1 square foot.

A square the side of which measures 1 yard is called a square yard.

If the side of a certain square is 1 yard long, how many feet long is it?

If you cut a square yard of brown paper into strips a foot wide, how many strips will you have?

How many square feet in each strip?

How many square feet in the three strips?

How many square feet, then, in a square yard?

9 square feet make 1 square yard.

. How many square inches in a square 4 inches on a side? 6 inches? 7 inches? 8 inches? 9 inches?

How many square feet in a square 2 feet on a side? 3 feet? 4 feet? 5 feet? 6 feet? 7 feet?

How many pecks in 8 quarts? in 24 quarts?
How many pecks in 16 quarts? in 32 quarts?
How many bushels in 8 pecks? in 12 peeks?
How many bushels in 4 peeks? in 16 pecks?

Add, and give the answers in bushels:

bushels.	peeks.	quarts.		bushels.	peeks.	quarts.
5	3	3		3	2	6
4	2	4		7	3	6
6	1	5		8	1	7
8	0	4		6	3	5

How many quarts in 2 pints? in 6 pints?
How many quarts in 4 pints? in 8 pints?
How many gallons in 8 quarts? in 12 quarts?

Add, and give the answers in gallons:

gallons.	quarts.	pints.		gallons.	quarts.	pints.
5	3	1		8	2	1
6	2	1		6	3	0
7	2	1		9	3	1
8	3	1		7	3	0

How many feet in 12 inches? in 24 inches? in
36 inches? in 48 inches? in 60 inches?
How many yards in 3 feet? in 6 feet? in 9 feet?
in 12 feet? in 21 feet? in 27 feet? in 36 feet?

Add, and give the answers in yards:

yards.	feet.	inches.		yards.	feet.	inches.
6	1	9		12	2	3
5	2	7		15	1	4
7	2	6		13	2	3
3	2	2		19	0	2

The coins of the United States are made of gold, silver, nickel, or bronze.

The **double-eagle,** the eagle, the **half-eagle,** and the **quarter-eagle** are made of **gold.**

Twenty dollars make a double-eagle.

Ten dollars make an eagle.

Five dollars make a half-eagle.

Two and one-half dollars make a quarter-eagle.

The **dollar,** the **half-dollar,** the **quarter-dollar,** and **ten**-cent piece are made of silver.

How many cents make a dollar?

One hundred cents make a dollar.

How many cents make a half-dollar?

How many cents make a quarter-dollar?

A ten-cent piece is often called a **dime.**

How many cents make a dime?

The **five**-cent piece is made of nickel.

A five-cent piece is often called a **nickel.**

How many cents make a nickel?

The **one-cent piece** is made of bronze.

A one-cent piece is often called a **penny.**

How many dimes make a half-dollar?

How many nickels make a quarter-dollar?

How many quarter-dollars make a half-dollar?

How many nickels make a half-dollar?

How many quarter-dollars make a dollar?

How many dimes make a dollar?

How many nickels make a dollar?

The sign $ is called the **dollar sign,** and is placed before the figures.

One dollar is written $1, or $1.00.

Eleven dollars and twenty-five cents is written $11.25.

The dot after the $11 in $11.25 means that the two figures on the right of it stand for cents, and the figures on the left of it stand for dollars.

The dot between the figures for dollars and the figures for cents is called the decimal point.

Read: $5.03; $7.27; $42.56; $12.23; $13.67; $67.53; $18.91; $98.01; $107.31; $121.02.

How many places do the cents occupy?

The cents always occupy two places.

Write in figures:

Three dollars and five cents.

Forty-five dollars and seventy-three cents.

Thirty-five dollars and sixty-seven cents.

Nineteen dollars and eighteen cents.

Eighty-nine dollars and ten cents.

One hundred five dollars and two cents.

One hundred seventeen dollars and one cent.

One hundred three dollars and three cents.

One hundred nine dollars and five cents.

One hundred one dollars and one cent.

Two hundred seventy dollars and nine cents.

Two hundred dollars and eight cents.

Three hundred dollars and twenty-five cents.

Two hundred dollars and fifty cents.

Add :

$2.03	$8.12	$12.12	$14.05	$30.03
3.04	7.32	13.13	11.10	20.02
3.21	5.13	21.21	31.32	40.01
5.51	6.41	32.32	23.50	50.50

$5.43	$9.34	$8.27	$11.17	$13.37
1.27	2.18	9.36	25.25	72.26
3.19	6.25	10.19	37.37	87.19

Subtract :

$7.45	$7.89	$8.59	$9.33	$36.55
− 5.03	− 4.63	− 5.26	− 7.29	− 28.00

$9.51	$5.65	$6.41	$6.73	$17.44
−3.28	− 1.27	− 2.38	− 1.09	− 8.36

Multiply :

$1.13	$2.24	$5.10	$8.12	$9.08
3	4	5	6	7

$11.07	$12.09	$9.07	$7.09	$6.08
8	9	9	7	8

Divide :

$16.08 by 2.	$12.24 by 6.	$56.56 by 8.
$12.24 by 2.	$24.12 by 6.	$64.08 by 8.
$18.36 by 3.	$35.35 by 7.	$54.54 by 9.
$24.12 by 4.	$49.42 by 7.	$81.09 by 9.
$25.05 by 5.	$56.56 by 7.	$63.72 by 9.

Ten mills make 1 cent.

What part of a cent is one mill? 2 mills? 3 mills? 5 mills? 7 mills? 10 mills?

Since 1 mill is 1 tenth of a cent, how many cents are twenty mills? 30 mills? 50 mills?

We write mills on the right of cents.

Two dollars 87 cents and 5 mills are written $2.875.

Thirty-seven cents and 5 mills are written $0.375.

Read: $3.607; $5.546; $18.364; $0.253.

Write in figures:

Seven dollars sixty cents and eight mills.

Eleven dollars seventy-five cents and five mills.

Twenty-one dollars two cents and two mills.

Ninety-nine cents and seven mills.

A ten-cent piece is often called a **dime**.

Ten dimes make a dollar.

What part of a dollar is 1 dime? 2 dimes? 3 dimes? 4 dimes? 5 dimes? 6 dimes? 10 dimes?

How many tenths of a dollar make the dollar?

How many tenths of a cent make the cent?

How many tenths of *any unit whatever* make **the whole unit?**

Tenths occupy one place, the first place to the right of the decimal point.

The number seven and three-tenths is written 7.3.

The number 6.5 is read six and five *tenths.*

The number 0.7 is read seven *tenths.*

Since 100 cents make a dollar, 1 cent is 1 **hundredth** of a dollar.

How many hundredths of a dollar are 2 cents? 3 cents? 5 cents? 10 cents? 25 cents? 50 cents?

How many tenths of a dollar are 10 cents?

How many **hundredths** of a dollar are 10 cents?

How many hundredths, then, make 1 tenth?

> 10 **hundredths** make 1 **tenth.**
> 10 **tenths** make 1 **unit.**

The number, three and five hundredths, is writ ten, 3.05. The number, two and sixty-four hundredths, is written, 2.64.

Hundredths always occupy two places.

Read: 5.08; 7.21; 10.54; 17.27; 65.65; 7.6; 6.07; 8.9; 8.09; 7.8; 7.08; 90.9; 90.09; 81.81.

Write in figures:

Five and five tenths.

Seventy-five and eighty-six hundredths.

Nine hundred one and nine hundredths.

Seventy-six and twenty-five hundredths.

Fifty-five and fifty hundredths.

How many hundredths are:

> 8 hundredths + 9 hundredths?
> 14 hundredths − 5 hundredths?
> 16 hundredths − 7 hundredths?

3 × 4 hundredths?	$\frac{1}{9}$ of 63 hundredths?
7 × 8 hundredths?	$\frac{1}{7}$ of 56 hundredths?
6 × 9 hundredths?	$\frac{1}{4}$ of 36 hundredths?

The number denoted by figures at the right of the decimal point is called a **decimal number,** or simply a **decimal.**

In adding or subtracting numbers containing decimals *we put the decimal point in the result directly under the column of decimal points in the given numbers.*

Add :

51.8	26.7	36.3	63.8
36.2	37.5	57.3	38.6
47.6	62.5	25.6	32.7
15.5	54.7	47.5	87.9

8.15	7.62	6.33	3.68
2.63	7.35	3.57	6.38
7.46	2.65	5.26	2.37
5.51	4.57	7.45	7.89

8.51	7.62	6.33	8.63
−2.36	−3.57	−3.75	−6.83

92.3	64.7	62.5	75.4
−35.7	−26.5	−45.7	−55.5

9.32	6.74	2.56	7.37
−7.25	−2.65	−1.19	−2.89

3.77	81.2	47.6	56.2
−1.98	−36.9	−28.7	−19.5

A farmer paid $160 for a horse and ¼ as much for a cow. How much did he pay for the cow?

A lady bought some blankets for $15 and some silk for $25. She gave ten-dollar bills in payment. How many bills did she give?

A boy bought a pair of boots for $4.25. He gave a five-dollar bill in payment. How much change did he receive?

A man earned in a week $19.50, and spent $12.25. How much did he save?

James earned $6.25, and his brother gave him enough to make $10. How much did his brother give him?

What will 9 barrels of flour cost at $6.10 a barrel?

What will 8 sheep cost at $6.10 apiece?

What will 5 hats cost at $3.10 apiece?

A lady bought a shawl for $11.50, and a hat for $8. She gave a twenty-dollar bill in payment. How much change did she receive?

Henry bought 3 pounds of beefsteak at 23 cents a pound, and gave a dollar-bill in payment. How much change did he receive?

At $0.50 a pound, how many pounds of Jersey butter can be bought for $2.50?

How many pounds of coffee at $0.30 a pound can be bought for $0.90?

At 8 cents a pound, how many pounds of rice can be bought for $0.56?

THE YEAR.

How many months make one year?

Twelve months make a year.

The names of the months in order are:

January, February, March, April, May, June, July, August, September, October, November, December.

The spring months are March, April, May.

The summer months are June, July, August.

The autumn months are September, October, November.

The winter months are December, January, February.

Spring, summer, autumn, winter, are called the **four seasons** of the year.

Thirty days have September, April, June, and November.

February has 28 days, and in leap years 29 days.

The other months have 31 days each.

Three hundred sixty-five days make a year.

Three hundred sixty-six days make a leap year.

When the date of the year can be divided by 4 without remainder, or in case the date ends in two zeros by 400, the year is a leap year.

Which of these years are leap years? 1800; 1860; 1872; 1890; 1893; 1892; 1900; 2000.

In a common year, how many days from the beginning of the year to February 15? to March 31? to April 7? to May 1? to June 14? to July 20?

THOUSANDS.

The number, 10 hundred, is called a thousand.

A thousand is written 1,000.

A thousand and one is written 1,001.

Ten thousand and ten is written 10,010.

One hundred twenty thousand four hundred is written 120,400.

How many thousands and how many ones in 7,632? 50,023? 41,701? 417,203? 500,230?

Write in figures and read all the numbers from 4,002 to 4,020; from 80,997 to 81,010; from 537,091 to 537,102; from 748,987 to 749,000.

Read: 5,430; 3,072; 1,010 ; 45,320; 70,045; 40,309; 36,008; 113,079; 273,002; 182,012; 811,200; 100,256; 500,005; 300,023; 608,300.

Write in figures:

Four thousand.   Three thousand seven.

Six thousand ten.   Five thousand fifteen.

Eight thousand three.

Nine thousand seven hundred.

Six thousand twenty-eight.

Seventy-four thousand six hundred.

Fifteen thousand five hundred.

Sixty-nine thousand thirty-two.

Seventy-three thousand five hundred forty-six.

Eight hundred thousand seven hundred five.

Ninety-six thousand eight hundred fifty-six.

Two hundred fifty thousand two hundred fifty.

Two hundred five thousand two hundred five.

MILLIONS.

When we write numbers which contain thousands and ones, we generally leave a little space after the last figure of the thousands, and put a comma in this space.   Thus, 236 347 is written 236,347.

This comma divides the figures into two **periods,** the period of thousands and the period of ones.

Forty-eight thousand and thirty-six sheep is written, 48,036 sheep.   Here we write 48 for the word *forty-eight;* then put a comma after the 8 for the word *thousand;* then 0, as there are no hundreds, and lastly, 36 for the word *thirty-six.*

The unit for the ones' period is 1 sheep.

The unit for the thousands' period is 1000 sheep.

The unit for the next higher period is a **million.**

A million is 1000 thousands, and is written

$$1,000,000.$$

The unit of any period is equal to 1000 units of the next lower period.

Three hundred million two hundred forty-six thousand five hundred dollars is written

$$\$300,246,500.$$

Here we put a comma after the 300 for the word *million,* and after the 246 for the word *thousand.*

The left-hand period may have one, two, or three figures, but every other **period** must have three **figures,** one figure for the hundreds, one for the tens, and one figure for the ones, of that period.

How many millions, thousands, and ones in 50,032,106 ? 41,107,106 ? 500,200,300 ?

Read :

32,027,020	316,106,207
100,370,200	70,000,035
275,701,050	170,202,305
75,017,500	28,028,280
57,207,005	202,170,503
10,987,278	111,798,827
65,371,954	210,007,500
87,250,520	120,052,250
54,054,540	540,504,054
95,720,027	905,059,950

Write in figures :

Thirty million, twenty-seven thousand, one hundred twenty dollars ?

Two hundred seven million, seven hundred twenty thousand, three hundred dollars.

Ninety-five million, fifty-nine thousand, one hundred sixty-six dollars.

Five hundred nine million, five hundred four thousand, five hundred forty dollars.

Twenty million, two hundred twenty thousand, three hundred sixty-four dollars.

Nineteen million, nineteen thousand, nine hundred nineteen dollars.

Thirty-seven million, three hundred thirty-seven thousand, seven hundred dollars.

Two hundred twenty million, three hundred thirty thousand, four hundred forty dollars.

THOUSANDTHS AND TEN-THOUSANDTHS.

If a unit is divided into **ten equal parts**, each part is called a **tenth** of the unit; if into a **hundred equal parts**, each part is called a **hundredth**; if into a **thousand equal parts**, each part is called a **thousandth**; if into **ten thousand equal parts**, each part is called a **ten-thousandth**.

NOTE. The Teacher should use the meter stick to show the decimal parts of a unit. The decimeters show the tenths, the centimeters the hundredths, and the millimeters the thousandths, of the meter.

Tenths occupy **one** decimal place 0.1
Hundredths occupy **two** decimal places 0.21
Thousandths occupy **three** decimal places . . . 0.213
Ten-thousandths occupy **four** decimal places . . 0.2134

The decimal 0.1 is read one *tenth ;* 0.21 twenty-one *hundredths ;* 0.213 two hundred thirteen *thousandths ;* 0.2134 twenty-one hundred thirty-four *ten-thousandths ;* 4.4045 is read four *and* four thousand forty-five *ten-thousandths.*

NOTE. In reading a number, part of which is integral and part decimal, pronounce *and* at the decimal point and omit it in all other places.

Read : 1.09 ; 23.023 ; 50.107 ; 7.0017 ; 7.0209 ; 5.5055 ; 2.3785 ; 15.0015 ; 6.2567.

Write in figures : two and five tenths ; two and five hundredths ; two and five thousandths ; two and five ten-thousandths ; two and twenty-five hundredths ; two and twenty-five thousandths ; two and twenty-five ten-thousandths ; two and two hundred twenty-five thousandths ; two and two hundred twenty-five ten-thousandths.

ADDITION.

To *test* the correctness of the work in addition, we add in a different order. The results should be the same. Thus, if we have added from the bottom to the top, we add from the top to the bottom.

1.	2.	3.	4.	5.	6.	7.
321	615	522	178	312	124	673
502	143	617	512	723	780	485
279	687	843	296	677	379	289

8.	9.	10.	11.	12.	13.
4321	3214	5423	8372	70.52	58.23
2751	5467	6543	543	53.84	1.92
6284	873	7654	7941	98.72	64.95
863	9124	6785	9078	8.76	8.67

Arrange and add, taking care to *have units of the same order stand in the same column.*

Decimals are easily arranged by taking care to have the decimal points stand in a vertical column.

14. 43,307 ; 96,812 ; 60,798 ; 21,121.

15. 83,654 ; 34,747 ; 58,659 ; 32,321.

16. 59.852 ; 41.664 ; 68.054 ; 90.594.

17. 10.5921 ; 27.3007 ; 31.9789 ; 2.563.

18. \$5.86 ; \$561.75 ; \$28.32 ; \$40.50.

19. 121,016 ; 167,404 ; 84,121 ; 66,456.

20. 90.0542 ; 32.8971 ; 55.674 ; 348.78.

21. 64.3372 ; 6.4337 ; 0.3723 ; 100.733.

22. 0.415 ; 70.634 ; 121.5007 ; 8.3467.

23. 8.0218 ; 15.101 ; 12.0031 ; 0.2256.

24. 121.0015 ; 100.37 ; 148.561 ; 1121.505.

25. \$5.86 ; \$8.78 ; \$11.89 ; \$12.58 ; \$95.37 ; \$59.88.

SLATE EXERCISES.

1. John Dix deposited in the Third National Bank of Boston $4321, and a week later $13,893. How much did he deposit in all?

2. The steamer Majestic made on four successive days 503, 504, 505 and 505 miles. How many miles did she make in the four days together?

3. In 1890 the population of New York was 1,513,501, of Brooklyn 804,377, and of Jersey City 162,317. What was the population of these three cities?

4. In 1890 St. Louis had 460,357 inhabitants, Boston had 447,720, Baltimore 432,095, and San Francisco 297,990. How many had these four cities together?

5. In 1890 Chicago had 1,098,576 inhabitants, Milwaukee 206,308, Minneapolis 164,738, St. Paul 133,156. How many had these four cities together?

6. In 1890 Philadelphia had 1,044,894 inhabitants, Pittsburgh 238,473, Alleghany 104,967, Scranton 83,450. What is the population of the four largest cities of Pennsylvania?

7. In 1890 Cincinnati had 296,309, Cleveland 261,546, Buffalo 255,543, Detroit 205,669. How many had these four cities together?

8. In 1890 Washington had 228,160, New Orleans 241,995, Louisville 161,005, and Richmond 80,838. Find the population of these four cities together.

SUBTRACTION.

To *test* the correctness of the work in subtraction, we add the subtrahend and the remainder. The sum should be equal to the minuend.

Subtract 427 from 736.

736
427
———
309

Beginning on the right, subtract 7 from 16, and write 9 below.

Afterwards subtract 2, not from 3, but from 2, and write 0 below. Then subtract 4 from 7, and write 3 below.

Subtract 7658 from 9000.

9000
7658
———
1342

Subtract 8 from 10, and write 2; then subtract 5, not from 10, but from 9, and write 4 ; again, subtract 6 from 9, and write 3 ; then subtract 7 from 8, and write 1.

Proof. Add 427
               309
              ———
               736

Proof. Add 7658
               1342
              ———
               9000

Subtract :

1. 873
   169
   ———

6. 3850
   1929
   ———

11. 60570
    48692
    ———

16. 462085
    345396
    ———

2. 679
   298
   ———

7. 5435
   1567
   ———

12. 20729
    17934
    ———

17. 701406
    243859
    ———

3. 700
   177
   ———

8. 5634
   5284
   ———

13. 32405
    21657
    ———

18. 740052
    698253
    ———

4. 901
   475
   ———

9. 9005
   6476
   ———

14. 20604
    11847
    ———

19. 402701
    317485
    ———

5. 506
   347
   ———

10. 3401
    2085
    ———

15. 60004
    28597
    ———

20. 400100
    375916
    ———

SUBTRACTION OF DECIMALS.

In the subtraction of decimals, make the number of decimal places in the minuend and subtrahend the same, annexing zeros if necessary.

Subtract 25.468 from 52.1253; and 2.1789 from 7.2.

OPERATION.	OPERATION.
52.1253	7.2000
25.4680	2.1789
26.6573	5.0211

Arrange so that the decimal point of the subtrahend shall be under that of the minuend, and subtract :

1. 0.85 − 0.79.
2. 1.76 − 0.98.
3. 2.729 − 1.836.
4. 5.482 − 3.176.
5. 2.354 − 2.287.
6. 3.826 − 3.719.
7. 5.902 − 3.678.
8. 5.77 − 4.888.
9. 9.62 − 3.765.
10. 8.42 − 5.661.
11. 7.23 − 6.562.
12. 9.02 − 7.163.
13. 4.31 − 3.425.
14. 1.27 − 1.198.
15. 1.46 − 0.955.
16. 13.2589 − 10.06.
17. 71.1002 − 52.387.
18. 11.2487 − 5.3579.
19. 10.9041 − 9.8765.
20. 17.3258 − 16.37.
21. 2.5 − 0.025.
22. 75 − 0.7575.
23. 1.52 − 1.0024.
24. 129.5 − 96.349.
25. 0.157 − 0.1547.
26. 752.8 − 4.9732.
27. 819.3 − 57.687.
28. 83.52 − 64.743.
29. 61.98 − 4.3554.
30. 6.716 − 0.8725.

SLATE EXERCISES.

1. Shakespeare was born in 1564 and died in 1616. How many years did he live?

2. Milton was born in 1608 and died in 1674. How many years did he live?

3. Daniel Webster died in 1852 at the age of 70. In what year was he born?

4. President Washington's first inaugural address contained 1300 words. His second inaugural address contained 134 words. How many more words did the first contain than the second?

5. President Lincoln's first inaugural address contained 3500 words. His second inaugural address contained 580 words. How many more words did the first contain than the second?

6. The population of Kansas City was 55,585 in 1880, and 132,416 in 1890. Find the increase.

7. The population of Denver was 35,629 in 1880, and 106,670 in 1890. Find the increase.

8. The population of Omaha was 30,518 in 1880, and 139,526 in 1890. Find the increase.

9. The number of silk looms in the United States in 1880 was 8474, and in 1890 the number was 22,569. Find the increase.

10. There are CL Psalms. James has read XCIX. How many more has he to read?

11. A woman bought groceries to the amount of $ 3.83. She gave a five-dollar bill in payment. How much change should she receive?

MULTIPLICATION.

If a product greater than 9 is obtained in multiplying, the figure for the *ones* only is written, and the *tens* are added to the following product.

$$
\begin{array}{r}
358 \\
4 \\
\hline
1432
\end{array}
$$

Thus, in the problem in the margin $4 \times 8 = 32$, we write the 2; then 4×5 tens $= 20$ tens, and to the 20 tens we add the 3 tens of the last product, obtaining 23 tens or 2 hundreds and three tens; we write the 3; then 4×3 hundreds $= 12$ hundreds, and to the 12 hundreds we add the 2 hundreds of the last product, obtaining 14 hundreds, which we write. The entire product is therefore 1432.

SLATE EXERCISES.

1. 2×3687.	18. 5×8267.	35. 4×29354.
2. 2×4783.	19. 6×6754.	36. 5×70528.
3. 3×2879.	20. 7×7854.	37. 6×56713.
4. 3×3657.	21. 7×9384.	38. 7×31567.
5. 5×1953.	22. 8×4337.	39. 8×37582.
6. 5×2849.	23. 3×9785.	40. 9×56014.
7. 4×3567.	24. 3×8694.	41. 9×34749.
8. 4×2586.	25. 7×2334.	42. 9×36927.
9. 5×6852.	26. 9×1682.	43. 9×73186.
10. 6×1376.	27. 5×9889.	44. 8×25839.
11. 6×5647.	28. 4×8977.	45. 7×98325.
12. 6×3124.	29. 6×9778.	46. 8×63578.
13. 3×8798.	30. 7×3879.	47. 9×67489.
14. 7×2342.	31. 9×3355.	48. 7×38697.
15. 8×4323.	32. 8×6675.	49. 9×48769.
16. 9×5215.	33. 7×8643.	50. 7×57009.
17. 4×7826.	34. 9×6854.	51. 8×99798.

If the multiplier has two or more figures :

We multiply by each figure separately, taking care to put the first figure of each product directly under the figure of the multiplier used in obtaining it; and add the products. Thus,

		Proof.	2046
7235			7235
2046			———
———			10230
43410			6138
28940			4092
14470			14322
———————			———————
14802810			14802810

The multiplicand and multiplier are called *factors* of the product. If either factor is 0, the product is 0. The product of two factors is not changed if the *order* of the factors is changed.

To *prove* multiplication, we change the order of the factors, and multiply again. The products should be the same in both cases.

Multiply :

1. 114 by 32.
2. 112 by 76.
3. 365 by 56.
4. 372 by 23.
5. 283 by 64.
6. 564 by 47.
7. 259 by 57.
8. 538 by 38.
9. 467 by 59.
10. 736 by 94.

11. 714 by 48.
12. 578 by 97.
13. 842 by 86.
14. 682 by 69.
15. 792 by 79.
16. 8763 by 407.
17. 8437 by 502.
18. 9872 by 603.
19. 7356 by 805.
20. 5983 by 704.

21. 3159 by 507.
22. 3819 by 206.
23. 8769 by 517.
24. 5731 by 475.
25. 8592 by 486.
26. 7069 by 908.
27. 5604 by 609.
28. 6789 by 789.
29. 4769 by 687.
30. 6897 by 976.

If the multiplier is 10, 100, 1000, etc., we obtain the product by annexing to the multiplicand as many zeros as there are in the multiplier.

Thus, 100 times 746 is 74,600.

In short, if one or both factors end in zeros, we multiply without regard to the zeros.

Then we annex to the product as many zeros as there are at the ends of the factors together. Thus,

To multiply 74,200 by 230, we first multiply 742 by 23, and obtain 17,066. To this number we annex 3 zeros, and get 17,066,000 for the true result.

Multiply :

1. 467   by 10.
2. 312   by 100.
3. 587   by 1000.
4. 6112   by 3000.
5. 7281   by 4000.
6. 8127   by 5000.
7. 43070 by 2000.
8. 43200 by 2340.
9. 56000 by 3480.
10. 50060 by 7000.
11. 50400 by 2080.
12. 47000 by 2070.
13. 504304 by 100.
14. 7120   by 7002.
15. 102039 by 112000.
16. 932600 by 184900.

17. If a man takes 180 steps a minute, how many steps will he take in an hour ?

18. If a man takes 2400 steps a mile, how many steps will he take in walking 20 miles ?

19. A cat has 18 toes. How many toes will 6000 cats have ?

20. At 60 cents a yard, what will be the cost of digging a drain 350 yards long ?

If one or both factors have decimal places:

We multiply without regard to the decimal point. Afterwards we point off in the product as many decimal places as there are decimal places in the two factors together. Thus:

Multiply 20.15 by 0.05.

$$
\begin{array}{r}
20.15 \\
0.05 \\
\hline
1.0075
\end{array}
$$

We multiply 20.15 by 0.05 and obtain 10075. As there are 2 decimal places in the multiplicand and 2 in the multiplier, we point off 4 decimal places in the product and have 1.0075, one and seventy-five *ten-thousandths*.

SLATE EXERCISES.

Multiply :

1. 0.541 by 444.
2. 0.853 by 232.
3. 3764 by 0.47.
4. 32.12 by 1.73.
5. 7860 by 46.8.
6. 0.623 by 373.
7. 763.2 by 8.65.
8. 68.42 by 75.5.
9. 8730 by 0.05.
10. 2.406 by 0.35.
11. 0.048 by 723.
12. 0.008 by 2.05.
13. 22.74 by 0.525.
14. 3792 by 0.024.
15. 0.715 by 141.5.
16. 466.4 by 45.06.
17. 3.417 by 1000.
18. 0.955 by 10000.
19. 6781 by 1.007.
20. 527.1 by 0.103.
21. 56.95 by 0.45.
22. 426.8 by 0.204.
23. 84.49 by 54.49.
24. 700.7 by 7.071.

SLATE EXERCISES.

1. A clock that strikes the hours, and 1 for the first quarter, 2 for the second and 3 for the third quarter, of each hour, strikes 300 times a day. How many times will it strike in a common year?

2. A clock that strikes the hours only, strikes 156 times in a day. How many times will it strike in a leap year?

3. If corn is $1.12 a bag, how much will 60 bags cost?

4. If coal is $5.75 a ton, how much will 17 tons cost?

5. If pine wood is $3.50 a cord, how much will 19 cords cost?

6. A farmer has 37 acres of corn worth on the average $27 an acre. What is the total value of his corn crop?

7. The earth moves in its orbit 19 miles a second. How many miles does it move in 1 minute?

8. If a bricklayer earns on the average $20.25 a week, how much will he earn in 28 weeks?

9. The lunar month is 29.53 days. How many days are there in 12 lunar months?

10. Sound travels at the rate of 1120 feet a second. Find the distance of a thunder-cloud when the thunder is heard 13 seconds after the lightning is seen.

11. A dealer sold 27 bushels of potatoes at 30 cents a peck. How much did he receive?

Divide 654 by 3.

Here $6 \div 3 = 2$, and as 6 is in the place of hundreds, we write 2 in the place of hundreds under the 6.

$$3) \overline{654} \\ \overline{218}$$

Then $5 \div 3 = 1$, with remainder 2.

We write the 1 in the place of tens, under the 5.

The remainder 2 is 2 tens or 20 ones, and 20 ones put with the 4 ones make 24 ones.

Then $24 \div 3 = 8$, and we write 8 in the place of ones, under the 4.

The quotient, therefore, is 2 hundreds, 1 ten, and 8 ones; that is, 218.

Divide 564 by 3.

Here $5 \div 3 = 1$, with remainder 2. We write the 1 in the place of hundreds, under the 5.

$$3) \overline{564} \\ \overline{188}$$

The remainder 2 is 2 hundreds, or 20 tens, and 20 tens put with 6 tens make 26 tens.

Then $26 \div 3 = 8$, with remainder 2.

We write the 8 in the place of tens, under the 6.

The remainder 2 is 2 tens, or 20 ones, and 20 ones put with 4 ones make 24 ones.

Then $24 \div 3 = 8$, and we write 8 in the place of ones, under the 4.

The quotient, therefore, is 1 hundred, 8 tens, and 8 ones; that is, 188.

Divide 765 by 9.

Since 7 will not contain 9, we take for the first partial dividend 76. Then $76 \div 9 = 8$ with remainder 4, and as 6, the last figure of this dividend, is in the place of tens, we write the quotient 8 in the place of the tens under the 6.

$$9) \overline{765} \\ \overline{85}$$

The remainder 4 is 4 tens or 40 ones, and 40 ones put with the 5 ones make 45 ones.

Then $45 \div 9 = 5$.

The quotient, therefore, is 8 tens and 5 ones; that is, 85.

Divide by 2 :

| 468 | 456 | 372 | 332 | 634 | 972 |
| 326 | 254 | 214 | 548 | 418 | 908 |

Divide by 3 :

| 354 | 365 | 624 | 484 | 408 | 798 |
| 444 | 235 | 651 | 790 | 891 | 976 |

Divide by 4 :

| 924 | 824 | 956 | 564 | 592 | 918 |
| 752 | 912 | 734 | 723 | 712 | 513 |

Divide by 5 :

| 510 | 520 | 640 | 770 | 590 | 745 |
| 665 | 735 | 560 | 880 | 620 | 825 |

Divide by 6 :

| 666 | 636 | 732 | 726 | 822 | 924 |
| 624 | 720 | 744 | 810 | 846 | 933 |

Divide by 7 :

| 728 | 784 | 812 | 861 | 910 | 945 |
| 742 | 797 | 805 | 875 | 931 | 952 |

Divide by 8 :

| 808 | 832 | 912 | 336 | 416 | 256 |
| 816 | 840 | 920 | 352 | 424 | 264 |

Divide by 9 :

| 927 | 945 | 405 | 378 | 288 | 135 |
| 936 | 918 | 396 | 387 | 297 | 225 |

SHORT DIVISION.

When the divisor is so small that the work can be performed mentally, the process is called **short division**.

Divide 63169 by 7.

7) 63169

9024 with rem. 1.

WORDING : 7 in 63, **9** ; in 1, **0** ; in 16 **2** ; in 29, **4**, with rem. 1.

EXPLANATION : Since 7 is not contained in 6, we take two figures 63 for the *first partial dividend*, and write the quotient 9 under the *right-hand figure* 3 of this partial dividend. 7 is not contained in 1, so 0 is written as the second figure of the quotient, and this 1, which is equal to 10 units of the next lower order of units, is joined to the 6, and makes 16 for the next partial dividend. Then 16 is divided by 7 ; the quotient is 2 and the remainder 2 ; the remainder 2 is equal to 20 of the next lower order of units, and with the 9 makes 29. Then 29 is divided by 7 ; the quotient is 4 and the remainder 1. Therefore the quotient is 9024, and the remainder is 1.

To *prove* division, we find the product of the divisor and quotient, and to this product add the remainder. The result should be equal to the dividend.

Proof.

9024
7
———
63168
1
———
63169

The product of the divisor and quotient is 63168.

To this product add the remainder 1, and the result is 63169, the same as the dividend.

Divide $54322 by $9.

$9) $54322

6035 with $7 rem.

In this example we are required to find the *number of times* we can take away $9 from $54322. The answer is 6035 *times,* with $7 over. The complete quotient may be written 6035⅞.

Divide $54322 by 9.

9) $54322

$6035 with $7 rem.

In this example we are required to divide $54322 into *nine equal parts*, and to find the *number of dollars* in each part. The answer is 6035 *dollars*, with $7 over. The answer may be written $6035⅞.

The last two examples illustrate the different meanings of division. *If the divisor and dividend refer to the same kind of units*, the quotient denotes the *number of times* the divisor must be taken to equal the dividend. If the divisor is *an abstract number* as 2, 3, 4, etc., the quotient denotes *a number of units of the same kind as the units of the dividend.*

Divide:

1. 434 by 2.
2. 876 by 3.
3. 596 by 4.
4. 432 by 4.
5. 180 by 5.
6. 715 by 5.
7. 875 by 5.
8. 618 by 6.
9. 324 by 6.
10. 819 by 7.
11. 847 by 7.
12. 920 by 8.
13. 904 by 8.
14. 945 by 9.
15. 621 by 9.
16. 513 by 2.
17. 707 by 3.
18. 845 by 4.
19. 901 by 5.
20. 862 by 6.
21. 872 by 7.
22. 907 by 9.

23. 5794 by 2.
24. 5874 by 3.
25. 5696 by 4.
26. 8975 by 5.
27. 3354 by 6.
28. 1176 by 7.
29. 8568 by 8.
30. 2943 by 9.
31. 3711 by 2.
32. 3226 by 3.
33. 8467 by 4.
34. 9573 by 5.
35. 6983 by 6.
36. 8659 by 7.
37. 4329 by 8.
38. 8256 by 9.
39. 5879 by 3.
40. 7361 by 9.
41. 6539 by 8.
42. 5396 by 7.
43. 9751 by 3.
44. 6857 by 7.

45. 95874 by 2.
46. 45873 by 3.
47. 46372 by 4.
48. 78295 by 5.
49. 66372 by 6.
50. 92582 by 7.
51. 87824 by 8.
52. 98172 by 9.
53. 78956 by 7.
54. 65978 by 8.
55. 76598 by 6.
56. 83621 by 3.
57. 86123 by 6.
58. 38612 by 9.
59. 12386 by 7.
60. 50080 by 8.
61. 65387 by 7.
62. 75429 by 5.
63. 31285 by 6.
64. 29514 by 9.
65. 65387 by 8.
66. 57148 by 3.

LONG DIVISION.

The process of Long Division is the same as that of Short Division, except that the work is written in full, and the quotient is written *over* the dividend.

Divide 31864 by 87.

The beginner will find it convenient to form a table of products of the divisor by the numbers 1, 2, 3, ..., as follows:

$1 \times 87 = 87$	$4 \times 87 = 348$	$7 \times 87 = 609$
$2 \times 87 = 174$	$5 \times 87 = 435$	$8 \times 87 = 696$
$3 \times 87 = 261$	$6 \times 87 = 522$	$9 \times 87 = 783$

As 87 is more than 31, it is necessary to take *three* figures of the dividend for the first partial dividend. Of the products in the table that do not exceed 318, the greatest is 261; that is, 3×87. Hence the first quotient figure is 3, and is written over the 8, the *right-hand figure* of the first partial dividend; then 261 is subtracted from 318. To the remainder 57, the next figure 6 of the dividend is annexed. Of the products that do not exceed 576, the greatest is 522; that is, 6×87. Hence 6 is the next figure of the quotient, and the next remainder is 54, to which the 4 of the dividend is annexed. Of the products that do not exceed 544, the greatest is 522; that is, 6×87. Hence the next figure of the quotient is 6, and the remainder 22. Therefore the quotient is 366, and the remainder 22.

OPERATION.

```
        366
87) 31864
    261
    ───
    576
    522
    ───
    544
    522
    ───
     22 rem.
```

After a little practice the operation of division can be performed without the aid of a table of products.

If at any step the product is greater than the partial dividend, the number denoted by the quotient-figure is too large and must be diminished; if the remainder is greater than the divisor, the number denoted by the quotient-figure is too small and must be increased.

Divide 1006078 by 247.

The first partial dividend is 1006. We find that 5×247 is 1235,
which is greater than 1006, and therefore 5
is too large. We try 4, and find that 4×247
is 988. We write the 4 over the 6, the
right-hand figure of the partial dividend,
and subtract the 988 from 1006. To the
remainder 18 we annex 0, the next figure
of the dividend, and have 180. Since 247
is not contained in 180, we write 0 for the
next figure of the quotient, and annex to
180 the next figure of the dividend, 7. The
next figure of the quotient is not 9, for

OPERATION.

$$
\begin{array}{r}
4073 \\
247)\overline{1006078} \\
988 \\
\hline
1807 \\
1729 \\
\hline
788 \\
741 \\
\hline
47 \text{ rem.}
\end{array}
$$

$9 \times 247 = 2223$, and is not 8, for $8 \times 247 = 1976$, and each of these products is greater than 1807. We try 7, and find the product to be 1729, which is less than 1807. The remainder obtained by subtracting 1729 from 1807 is 78, to which we annex the 8 of the dividend, and have 788. The next figure of the quotient is 3, and the product of 3×247 is 741. Subtracting 741 from 788 we get 47 for the remainder of the division. Hence the quotient is 4073, and the remainder 47.

Divide :

1. 5938 by 36.	13. 8757 by 67.	25. 8332 by 71.
2. 5743 by 37.	14. 9212 by 91.	26. 9888 by 93.
3. 9853 by 49.	15. 2786 by 22.	27. 7112 by 43.
4. 7369 by 52.	16. 3764 by 29.	28. 2931 by 19.
5. 9423 by 63.	17. 6753 by 57.	29. 9213 by 29.
6. 6578 by 74.	18. 9362 by 89.	30. 8778 by 55.
7. 6457 by 59.	19. 8579 by 73.	31. 61238 by 101.
8. 3579 by 21.	20. 8957 by 79.	32. 86123 by 201.
9. 7436 by 34.	21. 7319 by 53.	33. 38612 by 302.
10. 4589 by 42.	22. 8609 by 61.	34. 23816 by 205.
11. 5936 by 47.	23. 6891 by 31.	35. 12386 by 502.
12. 8372 by 65.	24. 3954 by 23.	36. 83216 by 603.

1.	98245 by 704.	**28.**	200836 by 897.
2.	59824 by 215.	**29.**	650734 by 635.
3.	45982 by 316.	**30.**	573206 by 753.
4.	82459 by 638.	**31.**	732065 by 537.
5.	93827 by 859.	**32.**	723540 by 871.
6.	96548 by 789.	**33.**	680023 by 997.
7.	84596 by 627.	**34.**	650734 by 736.
8.	23469 by 295.	**35.**	572036 by 853.
9.	24963 by 468.	**36.**	704532 by 973.
10.	59376 by 261.	**37.**	432960 by 187.
11.	56379 by 237.	**38.**	349062 by 259.
12.	79476 by 732.	**39.**	802365 by 795.
13.	67532 by 557.	**40.**	690409 by 389.
14.	70456 by 678.	**41.**	109370 by 167.
15.	80026 by 709.	**42.**	963047 by 398.
16.	72345 by 567.	**43.**	750431 by 578.
17.	90365 by 463.	**44.**	895047 by 757.
18.	78659 by 741.	**45.**	938704 by 198.
19.	94158 by 429.	**46.**	618543 by 4021.
20.	48519 by 229.	**47.**	816354 by 2008.
21.	67857 by 479.	**48.**	543168 by 4307.
22.	99321 by 912.	**49.**	604307 by 4803.
23.	79132 by 811.	**50.**	729718 by 5184.
24.	83742 by 566.	**51.**	542385 by 4978.
25.	650734 by 537.	**52.**	604730 by 4758.
26.	732065 by 631.	**53.**	817279 by 9814.
27.	704523 by 873.	**54.**	729718 by 4918.

ORAL EXERCISES.

1. If 3 cords of wood cost $ 9, what will 4 cords cost?

Note. Require the pupil to analyze this and similar problems by the **unitary method.** Thus, if 3 cords cost $9, 1 cord will cost ⅓ of $9, or $3 ; and 4 cords will cost 4×$3, or $ 12.

2. If 4 men can mow a field in 6 days, how many days will it take 3 men to mow the field ?

Analysis. If it takes 4 men 6 days to mow a field, it will take 1 man 4×6 days, or 24 days ; if it takes 1 man 24 days to mow a field, it will take 3 men ⅓ of 24 days, or 8 days.

3. Find the cost of 7 barrels of flour, if 8 barrels cost $ 40.

4. Find the cost of 12 oranges, if 5 oranges cost 15 cents.

5. What will 12 lambs cost, if 3 lambs cost $12 ?

6. If 12 men can dig a certain ditch in 6 days, how many men will be required to dig the ditch in 8 days ?

7. If 8 pounds of sugar cost 40 cents, how many cents will 11 pounds cost ?

8. If 3 tons of coal cost $ 21, how much will 8 tons cost ?

9. If 4 men can build a wall in 5 days, how many men will be required to build it in 4 days ?

10. If 3 yards of cloth are worth $ 6, how much are 7 yards worth ?

11. If 2 lamps cost $ 8, what will 5 lamps cost ?

12. If 9 yards of muslin cost 63 cents, what will 8 yards cost ?

13. If 8 men can do a piece of work in 9 days, how many days will it take 6 men to do it ?

14. How many pounds of butter at 20 cents a pound must be given for 2 pounds of tea at 60 cents a pound ?

Part IV.

LESSON 1.

DIVISION OF DECIMALS.

In Division, if the dividend and divisor are both multiplied or both divided by the same number, the quotient is not changed. Thus, $18 \div 6 = 3$, and (when both dividend and divisor are multiplied by 2) $36 \div 12 = 3$. Again (when both dividend and divisor are divided by 2), $9 \div 3 = 3$.

If the divisor is a whole number, and the dividend has decimals: *We divide as in whole numbers, but write the decimal point in the quotient as soon as the decimal point in the dividend is reached.*

Divide 1.29 by 3.

3)1.29   Since 3 is not contained in 1, we write 0 under the 1;
0.43   then the decimal point, and afterwards we continue, 3 in 12, 4; 3 in 9, 3. The quotient is 43 *hundredths*.

Divide:

1. 3.27 by 3.	8. 89.6 by 32.	15. 416.64 by 112.
2. 4.64 by 4.	9. 17.92 by 16.	16. 4089.8 by 121.
3. 5.75 by 5.	10. 313.6 by 14.	17. 17.161 by 131.
4. 16.24 by 7.	11. 375.7 by 17.	18. 380.48 by 232.
5. 18.66 by 6.	12. 709.5 by 15.	19. 140.36 by 116.
6. 18.48 by 8.	13. 42.12 by 18.	20. 140.30 by 115.
7. 28.17 by 9.	14. 8.489 by 13.	21. 2702.7 by 117.

163

If the divisor has decimals, and the dividend is a whole number : *We annex as many zeros to the dividend as there are decimal places in the divisor, and remove the decimal point from the divisor.*

Divide 129 by 0.2.

2)1290   Here we add 0 to the 129, making 1290, and divide by
  645   2 ; in other words, we multiply both dividend and divisor
         by 10.

Divide :

1. 129 by 0.3.	8. 132 by 0.33.	15. 121 by 0.11.
2. 122 by 0.4.	9. 625 by 2.5.	16. 132 by 0.12.
3. 136 by 0.5.	10. 603 by 1.5.	17. 169 by 0.13.
4. 174 by 0.6.	11. 165 by 3.3.	18. 196 by 1.4.
5. 161 by 0.7.	12. 282 by 4.7.	19. 256 by 0.16.
6. 128 by 0.8.	13. 318 by 5.3.	20. 324 by 1.8.
7. 117 by 0.9.	14. 648 by 7.2.	21. 585 by 6.5.

If both the divisor and dividend have decimals : *We remove the decimal point from the divisor, and move the decimal point in the dividend to the right as many places as there are decimals in the divisor.*

Divide 1.29 by 0.3.

3)12.9   Here we carry the decimal point in the dividend one
  4.3   place to the right, and remove it from the divisor. In other
        words, we multiply both dividend and divisor by 10.

Divide :

22. 12.9 by 0.3.	28. 3.24 by 0.9.	34. 0.96 by 0.2.
23. 12.4 by 0.4.	29. 13.2 by 0.3.	35. 0.33 by 0.3.
24. 13.5 by 0.5.	30. 2.01 by 0.5.	36. 1.98 by 0.9.
25. 1.86 by 0.6.	31. 1.28 by 0.4.	37. 17.6 by 0.8.
26. 1.61 by 0.7.	32. 17.4 by 0.6.	38. 15.5 by 0.05.
27. 12.8 by 0.8.	33. 1.82 by 0.7.	39. 12.6 by 0.09.

Divide 28.3696 by 1.49.

OPERATION.

$$
\begin{array}{r}
19.04 \\
149\overline{)\,2836.96} \\
149 \\
\overline{1346} \\
1341 \\
\overline{596} \\
596 \\
\end{array}
$$

Here the decimal point is removed from the divisor, and is moved two places to the right in the dividend ; in other words, both dividend and divisor are multiplied by 100.

Find the quotients of

1. $80.24 \div 8$.
2. $12.5664 \div 4$.
3. $1301.4 \div 241$.
4. $2647.08 \div 324$.
5. $9.215 \div 0.08$.
6. $664.56 \div 0.18$.
7. $132.6 \div 425$.
8. $7.48 \div 0.085$.
9. $0.748 \div 44$.
10. $2878.2 \div 369$.
11. $2.3328 \div 0.36$.
12. $52.5 \div 0.025$.
13. $1521 \div 11.7$.
14. $7236 \div 1.44$.
15. $67288 \div 64.7$.
16. $73807 \div 0.023$.

17. $300 \div 0.015$.
18. $32 \div 0.064$.
19. $2.88 \div 0.0024$.
20. $6.2 \div 0.0025$.
21. $65.1021 \div 3.207$.
22. $7704.256 \div 928$.
23. $506.016 \div 753$.
24. $1.9248 \div 0.008$.
25. $62825 \div 1.75$.
26. $700727 \div 0.029$.
27. $276.766 \div 37.1$.
28. $0.1024 \div 2.56$.
29. $1024 \div 25.6$.
30. $1292 \div 3.23$.
31. $906.5 \div 0.185$.
32. $0.4496 \div 11.24$.

SLATE EXERCISES.

1. A box contains 1416 eggs. How many dozen eggs are there in the box?

2. If 13 yards of velvet cost $97.50, what is the price of one yard?

3. If $38,057 are divided into 19 equal parts, how many dollars will there be in each part?

4. How many times is the sum of $17 contained in $2890?

5. There are 320 rods in a mile. How many miles are there in 9280 rods?

6. At $16.50 a ton, how many tons of hay can be bought for $280.50?

7. At $5.75 a ton, how many tons of coal can be bought for $103.50?

8. At 24 cents a dozen, how many dozen eggs can be bought for $61.44?

9. I bought 96 shares of railroad stock for $12,000. How much did the stock cost a share?

10. If a field produces 4905 bushels of corn, producing on the average 45 bushels to the acre, how many acres does the field contain?

11. At $10.50 a ton, how many tons of plaster can be bought for $65.625?

12. A man bought a barrel of sugar, weighing 232 pounds, for $12.76. How many cents a pound did he pay for the sugar?

13. When the price of Messina oranges is $2.75 a box, how many boxes can be bought for $77?

14. In how many hours will a cistern holding 4200 gallons be filled by a pipe that discharges into it 175 gallons an hour?

If the divisor is not contained in the dividend without a remainder, zeros may be annexed to the dividend, and the division continued.

Divide 0.39842 by 3.7164 to four decimal places.

OPERATION.

0.107.2 ·

37164) 3984.2
3716 4
───────
267800
260148
───────
76520
74328
───────
2192

If the divisor is a whole number, and ends in zeros. *We cut off the zeros from the divisor, and move the decimal point in the dividend as many places to the left (prefixing zeros if necessary), as there are zeros cut off.*

Divide 42.08 by 8000.

OPERATION.

8) 0.04208
───────
0.00526

Here the three zeros are cut off from the divisor, and the decimal point in the dividend is moved three places to the left. In other words, both divisor and dividend are divided by 1000.

SLATE EXERCISES.

Divide to four decimal places :

1. 5.8 by 4.79.
2. 7.34 by 2.3.
3. 16.28 by 0.67.
4. 54.87 by 0.39.
5. 2.86 by 349.
6. 8.6 by 3000.
7. 95 by 7000.
8. 89 by 6700.
9. 0.32 by 410.
10. 0.51 by 3700.

SLATE EXERCISES.

1. The production of pig-iron in the United States for the census year of 1890 was 9,579,779 tons, and 3,781,021 tons for the census year of 1880. Find the increase.

2. In 1880 Alabama produced 62,336 tons of pig-iron, and 890,432 tons in 1890. How many times the production of 1880 is the production of 1890?

3. The production of steel rails in the United States in 1880 was 741,475 tons, and 2,036,654 tons in 1890. Find the increase.

4. The value of wool manufactures in the United States for the census year of 1890 was $337,768,524; of cotton manufactures $267,981,724; of silk manufactures $87,298,454. Find the total value of the products of these three industries.

5. Find the difference in value between the wool and the cotton manufactures of the United States in 1890.

6. The total area devoted to the cultivation of cereals in the New England States in 1889 was 580,297 acres, and in 1879 the total area was 746,128 acres. Find the decrease.

7. In 1889 New Hampshire raised 988,806 bushels of Indian corn from 23,746 acres. Find to two places of decimals the average number of bushels per acre.

8. In 1889 Iowa raised 313,130,782 bushels of Indian corn from 7,585,522 acres. Find to the nearest bushel the average number of bushels per acre.

9. In 1889 the United States raised 468,321,424 bushels of wheat from 33,575,898 acres. Find to the nearest bushel the average number of bushels per acre.

COMPOUND QUANTITIES.

A quantity expressed in a *single unit* is called a **simple quantity**; but a quantity expressed in *different units* is called a **compound quantity**.

Thus, 20¼ pounds is a simple quantity, but 20 pounds 4 ounces is a compound quantity.

A unit of greater value or measure than another is said to be of a higher denomination than the other.

Thus, the dollar is of a higher denomination than the cent, the pound than the ounce, the yard than the inch, the hour than the minute.

The process of changing the *denomination* in which a quantity is expressed, without changing the *value* of the quantity is called **reduction**.

If the change is from a higher denomination to a lower, it is called **reduction descending**; if from a lower to a higher, it is called **reduction ascending**.

Thus, 1 yard = 36 inches is an example of reduction descending; and 24 inches = 2 feet is an example of reduction ascending.

LIQUID MEASURE.

Liquid Measure is used in measuring liquids, as water, milk, etc.

TABLE.

4 gills (gi.) = 1 pint (pt.).
2 pints   = 1 quart (qt.).
4 quarts   = 1 gallon (gal.).
Hence, 1 gal. = 4 qt. = 8 pt. = 32 gi.

31¼ gal.   = 1 barrel (bbl.).
63 gal.   = 1 hogshead.

NOTE. Casks holding from 28 gal. to 43 gal. are called barrels, and casks holding from 54 gal. to 63 gal. are called hogsheads. If we say, however, that a cistern holds 100 barrels, we mean barrels of 31¼ gal. each; or if we say that a cistern holds 100 hogsheads, we mean hogsheads of 63 gal. each.

Reduce 10 gallons 3 quarts 1 pint to pints.

gal. qt. pt.

10   3   1   10 gal. = 10 × 4 qt. = 40 qt., and 40 qt. with the 3 qt.
4   added are 43 qt.
―
43   43 qt. = 43 × 2 pt. = 86 pt., and 86 pt. with the 1 pt.
2   added are 87 pt.
―
87   87 pt. *Ans.*

1. Reduce 5 quarts 3 pints to pints.

2. Reduce 3 quarts 1 pint to pints.

3. Reduce 7 gallons 1 pint to pints.

4. Reduce 1 gallon 1 pint to gills.

5. Reduce 8 gallons 1 pint to pints.

6. Reduce 11 gallons 1 quart to pints.

7. Reduce 2 barrels to quarts.

8. Reduce 3 hogsheads to pints.

Reduce 129 pints to higher units.

2|129 pt.   129 pt. = $\frac{129}{2}$ qt. = 64 qt. and 1 pt. over.
4| 64 qt. . . . 1 pt.   64 qt. = $\frac{64}{4}$ gal. = 16 gal. and no qt. over.
   16 gal. . . 0 qt.

                           16 gal. 0 qt. 1 pt. *Ans.*

9. Reduce 229 pints to higher units.

10. Reduce 51 pints to higher units.

11. Reduce 365 pints to higher units.

12. Reduce 222 pints to higher units.

13. Reduce 1052 pints to higher units.

14. Reduce 1727 gills to higher units.

Add 4 gal. 3 qt. 1 pt. ; 11 gal. 1 qt. ; 3 qt. 1 pt. ; and 25 gal. 2 qt. 1 pt.

gal.	qt.	pt.
4	3	1
11	1	0
	3	1
25	2	1
42	2	1

Write the quantities so that units of the same name shall be in the same column.

The sum of the pints is 3. Divide the 3 pt. by 2 (2 pt. = 1 qt.). The result is 1 qt. and 1 pt. Write the 1 pt. under the column of pints.

The sum of the quarts, including 1 qt. from the 3 pt., is 10. Divide the 10 qt. by 4 (4 qt. = 1 gal.). The result is 2 gal. and 2 qt. Write the 2 qt. under the column of quarts, and add the 2 gal. to the gallons in the column of gallons.

42 gal. 2 qt. 1 pt. *Ans.*

From 4 gal. 2 qt. 1 pt. take 2 gal. 3 qt. 1 pt.

gal.	qt.	pt.
4	2	1
2	3	1
1	3	0

Since 1 pt. − 1 pt. is 0 pt., write 0 under the column of pints.

Since 3 qt. are more than 2 qt., take 1 gal. from the 4 gal., reduce it to quarts, and add them to the 2 qt., making 6 qt. Then, 6 qt. − 3 qt. = 3 qt. Write 3 under the column of quarts. Then, 3 gal. − 2 gal. = 1 gal.

1 gal. 3 qt. *Ans.*

Add :

1.			**2.**			**3.**		
gal.	qt.	pt.	gal.	qt.	pt.	gal.	qt.	pt.
3	1	1	21	3	$1\frac{1}{2}$	43	1	1
7	3	1	18	2	$1\frac{1}{2}$	27	3	1
8	3	1	7	2	1	31	3	$1\frac{1}{2}$

Find the difference between :

4.			**5.**			**6.**		
gal.	qt.	pt.	gal.	qt.	pt.	gal.	qt.	pt.
21	2	1	18	2	0	27	2	$1\frac{1}{2}$
7	3	1	7	2	1	17	3	1

7. From a barrel that held just 40 gal. and 2 qt. of vinegar there were drawn 19 gal. and 1 pt. How much vinegar was left in the barrel ?

Multiply 27 gal. 3 qt. 1 pt. by 5.

gal. qt. pt.   5 × 1 pt. = 5 pt. = 2 qt. 1 pt.   Write the 1 pt.
27   3   1   under the pints, and reserve the 2 qt. to be added tc
         5   the product of 5 × 3 qt.

139 1 1   5 × 3 qt. = 15 qt., and 15 qt. + 2 qt. = 17 qt.
              = 4 gal. 1 qt.

Write the 1 qt. under the quarts and add the 4 gal. to 5 × 27 gal.

                      139 gal. 1 qt. 1 pt. *Ans.*

Divide 113 gal. 2 qt. by 4.

gal.   qt.   pt.   The quotient from dividing 113 gal. by 4 is
4)113 2   0   28 gal., and the remainder is 1 gal.
   28   1   1   Reduce the 1 gal. to quarts, and add them to
              the 2 qt. The sum is 6 qt.

The quotient from dividing 6 qt. by 4 is 1 qt., and the remainder
is 2 qt.

Reduce the 2 qt. to pints, and we have 4 pt. Then 4 pt. ÷ 4
= 1 pt.

                    28 gal. 1 qt. 1 pt. *Ans.*

Divide 12 gal. 1 qt. by 3 qt. 1 pt.

        12 gal. 1 qt. = 49 qt. = 98 pt.
        3 qt. 1 pt.   = 7 pt.
     and 98 ÷ 7 = 14. *Ans.*

Multiply :

    1. 7 gal. 3 qt. 1 pt. by 9.
    2. 31 gal. 2 qt. by 7.
    3. 3 qt. 1 pt. 3 gi. by 8.

Divide :

    4. 126 gal. 3 qt. 1 pt. by 6.
    5. 110 gal. 1 qt. by 7.
    6. 131 gal. by 8.

NOTE. Methods precisely similar to the preceding are employed
for the reduction, addition, subtraction, multiplication, and division
of *all* compound quantities.

DRY MEASURE.

Dry Measure is used in measuring dry articles, as grain, seeds, fruit, vegetables.

TABLE.

2 pints (pt.) = 1 quart (qt.).
8 quarts   = 1 peck (pk.).
4 pecks   = 1 bushel (bu.).

Hence 1 bu. = 4 pk. = 32 qt.

NOTE 1. The gallon of liquid measure contains 231 cubic inches. Therefore the quart of liquid measure contains $57\frac{3}{4}$ cubic inches. The bushel of dry measure contains 2150.42 cubic inches. Therefore the quart of dry measure contains $67\frac{1}{5}$ cubic inches.

NOTE 2. In measuring grain, seeds, and small fruits, the measure must be *even* full. In measuring apples, potatoes, and other large articles, the measure must be *heaping* full.

1. Reduce 5 bu. 3 pk. 4 qt. to quarts.

2. Reduce 4056 pt. to higher denominations.

3. Multiply 7 bu. 2 pk. 7 qt. by 9.

4. Divide 25 bu. 3 pk. 2 qt. by 7.

5. How many 4-quart measures will 2 bu. 2 pk. 4 qt. fill ?

6. Divide 20 bu. 2 pk. by 8.

Add :

7.			8.			9.		
bu.	pk.	qt.	bu.	pk.	qt.	bu.	pk.	qt.
5	1	3	8	3	1	121	1	7
3	3	3	9	3	7	156	3	6
7	2	7	9	3	6	132	3	5

Subtract :

10.			11.			12.		
bu.	pk.	qt.	bu.	pk.	qt.	bu.	pk.	qt.
5	2	2	8	1	2	150	2	5
3	1	7	4	3	3	136	3	7

AVOIRDUPOIS WEIGHT.

Avoirdupois Weight is used in weighing all articles except gold, silver, and precious stones.

TABLE.

16 ounces (oz.)	= 1 pound (lb.).
2000 pounds	= 1 ton (t.).

The long ton is used in the United States Custom Houses and in wholesale transactions in iron and coal.

112 pounds Avoirdupois	= 1 long hundredweight (cwt.).
2240 pounds Avoirdupois	= 1 long ton.

1 pound Avoirdupois = 7000 grains.

NOTE. Many articles are sold by weight, as follows:

1 bu. of wheat or beans	= 60 lb.	1 bu. of potatoes	= 60 lb.
1 bu. of corn or rye	= 56 lb.	1 barrel of flour	= 196 lb.
1 bu. of corn or rye meal or cr'ked corn	= 50 lb.	1 barrel of beef or pork	= 200 lb.
		1 cask of lime	= 240 lb.
1 bu. of oats	= 32 lb.	1 quintal of fish	= 100 lb.
1 bu. of barley	= 48 lb.	1 stone of iron or lead	= 14 lb.
1 bu. of timothy seed	= 45 lb.	1 pig of iron or lead	= 300 lb.

1. Reduce 3 long tons 12 cwt. 110 lb. to pounds.

2. Reduce 87,956 lb. of coal to long tons.

3. Multiply 3 t. 1200 lb. of hay by 5.

4. Divide 8 t. 1500 lb. of hay by 7.

5. Add 1 t. 1326 lb., 1 t. 1560 lb., 1 t. 1728 lb.

6. From 2 t. 1015 lb. take 1 t. 515 lb.

7. From a firkin of butter containing 42 lb. there were sold 13 lb. 10 oz. How much was left?

8. At 23 cents a pound, what will 3.5 lb. of steak cost?

9. At \$15 a ton, what will 3.75 tons of hay cost?

TROY WEIGHT.

Troy Weight is used in weighing gold, silver, and precious stones.

TABLE.

24 grains (gr.)   = 1 pennyweight (dwt.).
20 pennyweights = 1 ounce (oz.).
12 ounces   = 1 pound (lb.).

The pound Troy contains 5760 grains.

1. How many more grains does a pound Avoirdupois contain than a pound Troy?

2. Reduce 8 oz. 12 dwt. to pennyweights.

3. Reduce 1760 dwt. to higher denominations.

4. How many grains are there in an ounce of silver?

5. From 1 lb. Troy take 5 oz. 5 dwt.

6. If 1 dwt. of silver is worth 4 cents, find the value of an ounce.

7. How many spoons weighing 1 oz. 5 dwt. each can be made from 30 oz. of silver?

8. How many table-spoons weighing 2 oz. 17 dwt. each can be made from 310 oz. 13 dwt. of silver?

9. Divide 373 oz. 2 dwt. by 7.

10. Multiply 27 oz. 13 dwt. by 6.

11. Add 11 oz. 11 dwt. 15 gr.; 7 oz. 12 dwt. 19 gr.; 10 oz. 13 dwt. 17 gr.

12. From 7 oz. 19 dwt. take 3 oz. 19 dwt.

NOTE. Apothecaries, in compounding medicines, use the following:

APOTHECARIES' MEASURE.

60 minims ($\mathfrak{m}$) = 1 dram ($\mathfrak{m}$ lx.).
8 drams   = 1 ounce (fl. drm. viij.).
16 ounces   = 1 pint (fl. oz. xvj.).

TIME MEASURE.

Time Measure is used in measuring duration.

TABLE.

60 seconds (sec.)	= 1 minute (min.).
60 minutes	= 1 hour (hr.).
24 hours	= 1 day (dy.).
7 days	= 1 week (wk.).
365 days (or 52 wk. 1 dy.)	= 1 common year (yr.).
366 days	= 1 leap-year.
100 years	= 1 century.

1. Reduce 3 dy. 11 hr. 32 min. to minutes.

2. Reduce 7 hr. 30 min. 50 sec. to seconds.

3. Reduce 20,400 min. to higher denominations.

4. Reduce 481,200 sec. to higher denominations.

5. From 3 yr. 15 dy. take 2 yr. 12 dy. 23 hr.

6. Divide 10 wk. 5 dy. 9 hr. by 9.

7. Multiply 2 dy. 7 hr. 15 min. by 8.

8. From 6 dy. 5 hr. 48 min. 43 sec. take 13 hr. 30 min. 40 sec.

9. Divide 31 dy. 2 hr. 54 min. by 7.

COUNTING.

PAPER.		VARIOUS.	
24 sheets	= 1 quire.	12 things	= 1 dozen.
20 quires	= 1 ream.	12 dozen	= 1 gross.
2 reams	= 1 bundle.	12 gross	= 1 great gross.
5 bundles	= 1 bale.	20 things	= 1 score.

How many sheets make a ream?

How many pens make a gross?

How many buttons make a great gross?

How many years are 3 score and ten?

LONG MEASURE.

Long Measure is used in measuring lines or distances.

TABLE.

12 inches (in.)	= 1 foot (ft.).
3 feet	= 1 yard (yd.).
5½ yards, or 16½ feet	= 1 rod (rd.).
320 rods	= 1 mile (mi.).

1 mi. = 320 rd. = 1760 yd. = 5280 ft.

NOTE. A line = $\frac{1}{12}$ in. ; a barleycorn = $\frac{1}{3}$ in. ; a hand (used in measuring the height of horses) = 4 in. ; a palm = 3 in. ; a span = 9 in. ; a cubit = 18 in. ; a military pace = 2½ ft. ; **a chain = 4 rd.** ; a link = $\frac{1}{100}$ chain ; a furlong = $\frac{1}{8}$ mi. ; a knot (used in navigation) = 6086 ft.; a league = 3 knots ; a fathom (used in measuring depths at sea) = 6 ft.; a cable length = 120 fathoms.

NOTE. Lengths measured by yards are generally expressed in yards and fractions of a yard; and distances of 160 rd. and 80 rd. are called *half-miles* and *quarter-miles* respectively.

Reduce 283 inches to higher denominations.

$$12\,|\,\underline{283}$$
$$3\,|\,\underline{23} \ldots 7$$
$$7 \ldots 2$$

7 yd. 2 ft. 7 in. *Ans.*

Reduce 328 yards to rods.

$$5\tfrac{1}{2}\,|\,328$$
$$2$$
$$11\,|\,\underline{656} \text{ half-yards.}$$
$$59 \ldots 7 \text{ half-yards.}$$

Since it takes 5½ yards, or 11 **half-yards,** to make a rod, reduce the 328 yards to *half-yards* and divide by 11. The quotient is 59 rods, and the remainder is 7 half-yards. The 7 half-yards are equal to 3½ yards.

59 rds. 3½ yds. *Ans.*

What part of a yard are 9 in.? 18 in.?
What part of a mile are 160 rd.? 80 rd.?
How many yards in 2 rd.? in 3 rd.? in 4 rd.?
How many feet in 2 rd.? in 4 rd.? in 6 rd.?

1. Change 5 yd. 2 ft. 7 in. to inches.

2. Change 2 yd. 2 ft. 4 in. to inches.

3. Change 2 mi. 268 rd. to rods.

4. Change 16 mi. 181 rd. to rods.

5. Change 15,840 ft. to miles.

6. Change 935 yd. to rods.

7. Change 720 rd. to miles.

8. Change 19,360 yd. to miles.

Add:

9.			10.			11.		
yd.	ft.	in.	yd.	ft.	in.	mi.	rd.	yd.
13	1	5	27	4	1	15	25	5
28	2	7	14	3	2	3	27	3
5	2	11	14	3	2	12	36	2

12.			13.			14.		
mi.	rd.	ft.	mi.	rd.	ft.	rd.	ft.	in.
13	35	15	7	140	10	170	8	9
11	57	11	5	230	12	115	11	11
10	85	13	3	275	5	130	14	8
5	96	8	1	255	11	175	13	7

*Find the difference between:

15.			16.			17.		
yd.	ft.	in.	yd.	ft.	in.	mi.	rd.	ft.
14	1	4	22	0	0	23	76	1
3	1	5	3	2	6	16	238	15

18.			19.			20.		
mi.	rd.	ft.	mi.	rd.	yd.	mi.	rd.	yd.
17	125	1	7	0	0	13	33	2
8	257	14	3	255	1	4	0	3½

21. Multiply 15 yd. 1 ft. 9 in. by 11.

22. Multiply 21 rd. 4 yd. 2 ft. by 13.

PERIMETERS.

The **Perimeter** of any surface bounded by straight lines is the sum of the lengths of the bounding lines.

A **Rectangle** is a flat surface with four straight sides and four square corners.

If the four sides are equal, the rectangle is called a **Square.**

Find the perimeter of :

RECTANGLE.

1. A rectangular floor 15 ft. by 15 ft.
2. A rectangular ceiling 22 ft. by 20 ft.
3. A rectangular room 16 ft. by 18 ft.
4. A rectangular room 24 ft. by 21 ft.

A **Circle** is a flat surface bounded by a curved line called the **Circumference,** all points of which are equally distant from a point within called the **Centre.**

The length of the circumference of a circle is found by multiplying the length of the diameter by 22 and dividing the result by 7.

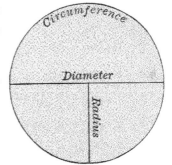

Find the length of the circumference of a circle :

5. If the length of the diameter is 21 in.; 28 in.; 7 ft.

The length of the diameter of a circle is found by multiplying the length of the circumference by 7 and dividing the result by 22.

Find the length of the diameter of a circle :

6. If the length of the circumference is 11 in.

7. If the length of the circumference is 2 ft. 9 in.

LESSON 18.

SQUARE MEASURE.

Square Measure is used in measuring surfaces.

The units of square measure are squares having *units of length* for the lengths of their sides.

TABLE.

144 square inches (sq. in.) = 1 square foot (sq. ft.).
9 square feet = 1 square yard (sq. yd.).
$30\frac{1}{4}$ square yards, or $\}$ $272\frac{1}{4}$ square feet $\}$ = 1 square rod (sq. rd.).
160 square rods, or $\}$ 10 square chains $\}$ = 1 acre (A.).
640 acres = 1 square mile (sq. mi.).

Hence, 1 A. = 160 sq. rd. = 4840 sq. yd. = 43,560 sq. ft.

A square of flooring or roofing = 100 sq. ft.
A section of land = 1 mile square.
A township = 36 sq. mi.

The units of square measure are obtained by squaring the units of long measure. Thus,

$$144 = 12^2 ; \quad 9 = 3^2 ; \quad 30\frac{1}{4} = (5\frac{1}{2})^2 ; \quad 272\frac{1}{4} = (16\frac{1}{2})^2.$$
12^2 is read the square of 12 and means 12×12.

1. Reduce 507 sq. yd. 7 sq. ft. to square feet.

2. Reduce 50 sq. chains to acres.

3. Reduce 3 A. 90 sq. rd. to square rods.

4. Reduce 44,996 sq. in. to square feet.

5. Reduce 67,760 sq. yd. to acres.

6. Reduce 85,316 sq. rd. to acres.

7. Add: 3 A. 116 sq. rd.; 2 A. 120 sq. rd.; 5 A. 119 sq. rd.; 1 A. 40 sq. rd.

8. From 13 sq. yd. 7 sq. ft. 12 sq. in. take 3 sq. yd. 8 sq. ft. 136 sq. in.

9. Multiply 2 A. 20 sq. rd. by 9.

AREAS.

The **area** of any surface is the **number of units of area** the given surface contains.

The **unit of area** is a square, the side of which is some given *unit of length.*

Find the area of a rectangle 2 ft. 3 in. by 1 ft. 8 in.:

2 ft. 3 in. = 24 in. + 3 in. = 27 in.
1 ft. 8 in. = 12 in. + 8 in. = 20 in.

Therefore the area required is 20 × 27 = 540 sq. in.

Hence, in finding the area of a rectangle :

We express the length and breadth in units of the same denomination, and multiply the number of units in the length by the number of units in the breadth ; this product will be the number of square units of that denomination.

Find the area of a rectangle:

1. 8 in. by 5 in.   **4.** 11 in. by 10 in.   **7.** 3 ft. by 2 ft.
2. 9 in. by 6 in.   **5.** 15 in. by 6 in.   **8.** 4 ft. by 2 ft.
3. 8 in. by 7 in.   **6.** 16 in. by 4 in.   **9.** 8 ft. by 2 ft.

10. How many square feet in a floor 21 ft. by 20 ft.?

11. How many square feet in a floor 18 ft. by 15 ft.?

12. How many square feet in a blackboard 12 ft. long, and 3 ft. wide?

13. How many square yards in a roll of wall-paper $\frac{1}{2}$ yd. wide and 8 yd. long?

14. Find the number of square yards in a house-lot 87 ft. front and 102 ft. deep.

15. Find the number of square rods in a house-lot 8 rods front and 10 rods deep.

16. Find the total area of the *four walls* of a room 18 ft. long, 15 ft. wide, and 9 ft. high.

1. Find the total area of the four walls and the ceiling of a room 16 ft. long, 15 ft. wide, and 10 ft. high.

2. Find the total area in square yards of the ceiling of a room 18 ft. long, and 15 ft. wide.

3. Find the area in square yards of the four walls of a room 19 ft. long, 17 ft. wide, and 9 ft. high.

4. Find the number of acres in a field 40 rods square.

5. Find the number of square yards in a flower bed that is 12 ft. square.

6. Find the number of square yards in a poppy bed that is 24 ft. long, and 12 ft. wide.

7. Find the number of square inches in the surface of a slate 8 in. by 14 in.

8. Find the number of square inches in the surface of a crayon-box 7 in. by 4 in. by 3 in.

9. Find the number of square feet in the surface of a cube 3 ft. by 3 ft. by 3 ft.

The area of a circle is found by multiplying the area of the square on its radius by 22 **and dividing the result by** 7.

Find the area of a circle:

10. If the length of the radius is 10 in.; 16 in.; 20 in.

11. If the length of the radius is 1 ft. 4 in.; 1 ft. 6 in.

12. If the length of the diameter is 1 ft. 10 in.; 2 ft. 4 in.

13. If the length of the diameter is 2 ft. 6 in.; 3 ft. 4 in.; 3 ft. 8 in.; 3 ft. 10 in.

14. If the length of the diameter is 4 ft. 2 in.

CARPETING FLOORS.

In carpeting floors, decide whether the strips shall run lengthwise of the room or across it, and find the number of strips required by dividing the width of the room by the width of the carpet, if the strips are to run lengthwise of the room; and the length of the room by the width of the carpet, if the strips are to run across it. A fraction of a width of carpeting required is reckoned a full width, and enough is **turned under** to make the carpet fit the room.

The number of yards in the length of the strip required multiplied by the number of strips will give the number of yards of carpeting required.

In determining the length of the strip, *allowance must be made for matching the patterns.*

Ex. Find the number of strips of carpeting 27 in. wide required for a room 18 ft. by 17 ft., if the strips run lengthwise.

SOLUTION.

$$17 \text{ ft.} = 17 \times 12 \text{ in.} = 204 \text{ in.}$$
$$204 \text{ in.} \div 27 \text{ in.} = 7\tfrac{15}{27}.$$

Therefore 8 strips are required.

1. Find the number of strips of carpeting 1 yd. wide required for a room 17 ft. by 15 ft., if the strips run lengthwise.

2. Find the number of strips of carpeting 27 in. wide required for a room 20 ft. by 22 ft. 6 in., if the strips run across the room.

3. Find the number of yards of carpeting 1 yd. wide required for a room 17 ft. 6 in. by 17 ft., if the strips run lengthwise. What width will be turned under?

PAPERING ROOMS.

Wall-paper is made in strips 18 in. wide. Single rolls are 8 yards long, and double rolls are 16 yards long.

To find the number of rolls required to paper a room of not more than eight feet from baseboard to border.

We find the number of feet in the perimeter of the room, omitting the width of the doors and windows; and allow a double roll, or two single rolls, for every 7 feet of the perimeter.

Find the number of rolls required for a room of ordinary height, 17 ft. by 15 ft., having 1 door and 3 windows each 4 ft. wide.

Perimeter of the room $= 2 \times 17$ ft. $+ 2 \times 15$ ft. $= 64$ ft.
Width of door and windows $\qquad = 16$ ft.
Deducting, we have $\qquad \overline{48 \text{ ft.}}$

$$48 \div 7 = 6\tfrac{6}{7}.$$

Ans. 7 double rolls.

1. How many double rolls of paper will be required for a room 20 ft. by 18 ft., with 2 doors and 3 windows, each 4 ft. wide?

2. Find the cost of paper at 50 cents a single roll for a room 21 ft. by 19 ft., with 2 doors and 4 windows, each 4 ft. 2 in. wide.

3. Find the cost of paper at 30 cents a single roll for a room 16 ft. by 15 ft., with 1 door and 2 windows, each 4 ft. wide.

4. How many double rolls of paper will be required for a room, if the perimeter is 68 ft. after allowance is made for doors and windows?

5. How many double rolls of paper will be required for a room, if the perimeter is 60 ft. after allowance is made for doors and windows?

BOARD MEASURE.

To find the measure of boards one inch thick or less :

We express the length in feet and the width in inches, and divide the product of these two numbers by 12 ; the quotient is the number of feet board measure.

How many feet in a board 18 ft. long, 16 in. wide ?

$$\frac{\overset{3}{\cancel{18}} \times \overset{8}{\cancel{16}}}{\cancel{12}} = 24. \qquad 24 \text{ ft. } Ans.$$

To find the measure of boards more than one inch thick, and of other squared lumber :

We express the length in feet, the width and the thickness in inches, and divide the product of these three numbers by 12 ; the quotient is the number of feet board measure.

Thus, the number of feet board measure in a plank 12 ft. long, 15 in. wide, and 3 in. thick is $\dfrac{12 \times 15 \times 3}{12} = 45$ ft.

How many feet are there in an inch board :

1. 16 ft. long, 12 in. wide ?   **4.** 20 ft. long, 10 in. wide ?
2. 18 ft. long, 16 in. wide ?   **5.** 18 ft. long, 8 in. wide ?
3. 12 ft. long, 14 in. wide ?   **6.** 15 ft. long, 12 in. wide ?

How many feet are there in a plank :

7. 20 ft. long, 18 in. wide, and 3 in. thick ?
8. 12 ft. long, 16 in. wide, and 2 in. thick ?
9. 12 ft. long, 10 in. wide, and 4 in. thick ?
10. How many feet are there in a stick of timber 16 ft. long, 10 in. wide, and 10 in. thick ?
11. How many feet are there in a joist 12 ft. long, 4 in. wide, and 4 in. thick ?
12. How many feet are there in a floor timber 21 ft. long, 10 in. wide, and 2 in. thick ?

CUBIC MEASURE.

Cubic Measure is used in measuring solids.

The units of cubic measure are cubes having units of length for the lengths of their edges.

TABLE.

1728 cubic inches (cu. in.) = 1 cubic foot (cu. ft.).
27 cubic feet   = 1 cubic yard (cu. yd.).

The units of cubic measure are cubes of the units of long measure. Thus, $1728 = 12^3$; $27 = 3^3$.

12^3 is read the cube of 12 and means $12 \times 12 \times 12$.

WOOD MEASURE.

TABLE.

16 cubic feet = 1 cord foot (cd. ft.).
8 cord feet = 1 cord (cd.).
Therefore, 128 cubic feet = 1 cord.

1. Reduce 13 cu. yd. 21 cu. ft. to cubic feet.

2. Reduce 600 cu. ft. to cubic yards.

3. From 58 cu. yd. 24 cu. ft. take 34 cu. yd. 26 cu. ft.

4. Multiply 13 cu. yd. 13 cu. ft. by 13.

5. Divide 17 cu. yd. 14 cu. ft. by 11.

6. Add: 34 cu. yd. 11 cu. ft.; 13 cu. yd. 10 cu. ft.; 17 cu. yd. 4 cu. ft.

7. How many cords of wood in 1280 cu. ft.?

8. How many cords of wood in a pile 42 ft. long, 8 ft. wide, and 6 ft. high?

NOTE. Divide the product of the *numbers* expressing the length, width, and height by 128.

9. How many cubic yards in a cord of wood?

10. At $4 a cord, find the value of a pile of wood 18 ft. long, 4 ft. wide, and 4 ft. high.

VÓLUMES.

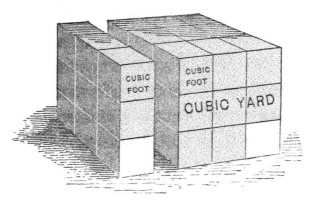

The **volume** of any solid is the **number of units of volume** the given solid contains.

The **unit of volume** is a cube, the edge of which is some given *unit of length.*

A **rectangular solid** is a solid bounded by six rectangles.

In **finding the volume of a rectangular solid:**

We express the length, breadth, and thickness in units of the same denomination; then we multiply the number of units in the length by the number in the breadth, and this product by the number in the thickness; the result will be the number of cubic units of that denomination.

Find the volume of a rectangular solid:

1. 8 in. × 4 in. × 3 in.   **4.** 7 in. × 3 in. × 4 in.

2. 4 in. × 4 in. × 3 in.   **5.** 10 in. × 8 in. × 4 in.

3. 4 in. × 4 in. × 4 in.   **6.** 3 in. × 3 in. × 3 in.

7. Find the number of cubic feet in a stick of square timber 30 ft. long, 1 ft. square at the end.

8. Find the number of cubic yards in an excavation for a cellar 42 ft. by 33 ft. by 9 ft.

9. Find the number of cubic yards in an excavation for a cellar 33 ft. by 24 ft. by 9 ft.

BILLS.

A **Bill of Goods** is a written statement of goods sold, and of payments, if any, received for them.

A **Bill of Services** is a written statement of services rendered, or of labor performed.

A **Statement of Account** is a statement of the sum due according to the accounts already rendered. Thus,

Mr. Jones,   To BROWN & CO., Dr.

June 1	*To Account rendered*	*$60*

The **Creditor** is the person who sells the goods, or who performs the labor.

The **Debtor** is the person who buys the goods, or who pays for the services rendered.

The **Debit Side of the Account** consists of the items due to the person who renders the account.

The **Credit Side** consists of the amounts received by the person who renders the account.

The **Balance of an Account** is the difference between the amounts of the Debit and Credit Sides.

NOTE When a bill is paid, it should be receipted by writing at the bottom of the bill the date of payment and the words *Received payment*, and under these words the creditor should sign his name and deliver the receipt to the debtor.

If a clerk has authority to sign his employer's name, he should write under his employer's name his own name preceded by the word **by** or **per.**

RECEIPTED BILLS OF GOODS.

Boston, June 1, 1893.

Mr. Robert Thomson,

Bought of CHARLES EDMONDS.

1893						
May	15	10 lb. Coffee	@ 35 ¢	$3	50	
		50 lb. Sugar	@ 5 ¢	2	50	
		2 lb. Tea	@ 5 ¢	1	30	
		28 lb. Butter	@ 62 ¢	8	96	
				16	26	

June 1, 1893.   Received payment,

Charles Edmonds.

Exeter, June 1, 1893.

James York,

To KELLY & GARDNER, Dr.

1893								
Mar.	8	To 2 gal. Molasses	@ 55¢	$1	10	$	
		To 2 bbl. of Flour	@ $5.75	11	50	12	60
Apr.	5	To 15 lb. Rice	@ 9¢	1	35		
		To 25 lb. Butter	@ 33¢	8	25	9	60
						22	20
		Cr.					
Mar.	8	By 2 cords Birches	@ $4.50¢	9	00		
		By 3 bu. Potatoes	@ 65¢	1	95	10	95
		Balance due				11	25

June 1, 1893.   Received payment,

Kelly & Gardner,

By James Staples.

Make out bills, and receipt for them the first day of the month that follows the purchase :

1. *Mr. Leonard Smith,*
Bought of JOHN THOMPSON & CO.

1893				
Feb.	7	9 lb. Ham	@ 15¢	
		18 " Steak	@ 25¢	
"	13	15 " Mutton	@ 16¢	
		11 " Veal	@ 11¢	

2. *James Coffin,*   To HOWARD MANSUR, Dr.

1893				
May	3	25 lb. Codfish	@ 9¢	
"	10	30 " Bacon	@ 14¢	
"	18	10 " Coffee	@ 35¢	
"	25	2 bbl. Flour	@ $5.75	

3. *John Marshall,*   To ROBERT STUART, Dr.

1893				
Mar.	8	12 doz. Eggs	@ 26¢	
		17 lb. Butter	@ 32¢	
"	15	34 " Cheese	@ 8¢	
		16 bu. Potatoes	@ 85¢	

4.

1893				
Apr.	5	27 bags Whole Corn	@ $1.12	
		30 " Meal	@ 1.00	
"	12	60 " Oats	@ 0.65	
		7 tons Hay	@ 17.00	
'	19	Cr.		
		By Cash	$200	

COMMON FRACTIONS.

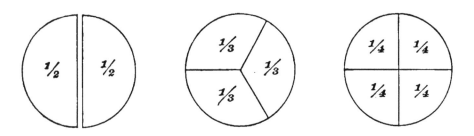

If a circle is divided into 2 equal parts,

What part of the whole circle is each of these parts?
What part of the whole circle are 2 of these parts?

If a circle is divided into 3 equal parts,

What part of the whole circle is each of these parts?
What part of the whole circle are 2 of these parts?
What part of the whole circle are 3 of these parts?

If a circle is divided into 4 equal parts,

What part of the whole circle is each of these parts?
What part of the whole circle are 2 of these parts?
What part of the whole circle are 3 of these parts?
What part of the whole circle are 4 of these parts?
How many *halves* of a unit make the whole unit?
How many *thirds* of a unit make the whole unit?
How many *fourths* of a unit make the whole unit?

What is the *name* of one of the parts of a unit,

When the unit is divided into *two equal parts?*
When the unit is divided into *three equal parts?*
When the unit is divided into *four equal parts?*
Which is larger $\frac{1}{2}$ of a circle or $\frac{1}{3}$ of the circle?
Which is larger $\frac{1}{2}$ of a circle or $\frac{1}{4}$ of the circle?
Which is larger $\frac{1}{3}$ of a circle or $\frac{1}{4}$ of the circle?

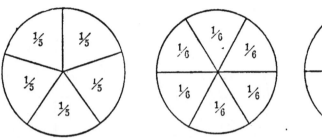
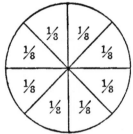

If a circle is divided into 5 equal parts,

What part of the circle is each of these parts?

What part of the circle are 2 of these parts? 3 of these parts? 4 of these parts? 5 of these parts?

If a circle is divided into 6 equal parts,

What part of the circle is each of these parts?

What part of the circle are 2 of these parts? 3 of these parts? 4 of these parts? 5 of these parts? 6 of these parts?

If a circle is divided into 8 equal parts,

What part of the circle is each of these parts?

What part of the circle are 2 of these parts? 4 of these parts? 6 of these parts? 7 of these parts? 8 of these parts?

How many *fifths* of a unit make the whole unit?

How many *sixths?* how many *sevenths?* how many *eighths?* how many *tenths?* how many *twelfths?* how many *sixteenths?*

What is the *name* of one of the parts of a unit, when the unit is divided into 5 equal parts? into 6 equal parts? into 8 equal parts? into 10 equal parts? into 12 equal parts?

Which is larger $\frac{1}{5}$ of a unit or $\frac{1}{6}$ of the unit? $\frac{1}{6}$ of a unit or $\frac{1}{8}$ of the unit? $\frac{1}{10}$ of a unit or $\frac{1}{12}$ of the unit.

Any standard used in counting or in measuring is called a unit.

In 3 quarters of a yard the **unit** is a *quarter of a yard.* But a quarter of a yard is a *fractional part* of the whole **unit,** a yard.

A unit which is a fractional part of another unit is called a **fractional unit,** and the unit of which it is a part is called **its whole unit.**

Numbers that count whole units are called **whole numbers.** Numbers that count fractional units are called **fractional numbers, or fractions.**

NOTE. The Teacher must explain that the words *whole* and *fractional*, though applied to numbers, refer only to the units counted by the numbers.

Name the fractional unit and the integral unit in:

3 quarters of an inch.	1 half of an hour.
4 fifths of a pound.	6 sevenths of a week.
2 thirds of a yard.	5 twelfths of a foot.
3 eighths of a bushel.	3 sixteenths of a ton.
9 tenths of a dollar.	5 sixths of an acre.

Express:

$\frac{1}{2}$ of a yard in inches.	$\frac{1}{2}$ of a pound in ounces.
$\frac{2}{3}$ of a yard in inches.	$\frac{3}{4}$ of a pound in ounces.
$\frac{3}{4}$ of a yard in inches.	$\frac{3}{8}$ of a pound in ounces.
$\frac{5}{6}$ of a yard in inches.	$\frac{5}{8}$ of a pound in ounces.

Every common fraction is written in figures by means of two numbers, which are called the **terms** of the fraction.

One of these gives the *name* of the parts, and is called the **denominator ;** and the other gives the *number* of the parts taken, and is called the **numerator.**

To write a common fraction, write the denominator under the numerator with a line between them.

To write 5 *sevenths*, for example, we write the numerator 5, draw a line under it, and under the line we write the denominator 7 ; thus, $\frac{5}{7}$.

To read a common fraction, read the numerator and then the denominator.

Thus, $\frac{2}{3}$, $\frac{1}{2}$, $\frac{3}{5}$, $\frac{7}{8}$, $\frac{9}{11}$, are read two-thirds, one-half, three-fifths, seven-eighths, nine-elevenths. $\frac{3}{4}$ is read three-fourths or three-quarters.

Write in figures :

one third.	seven twentieths.
one quarter.	thirteen twenty-fifths.
two fifths.	five sevenths.
five sixths.	nine thirteenths.
five eighths.	eleven twelfths.
seven twelfths.	four twenty-firsts.
three sixteenths.	seventeen eighteenths.
nine fourteenths.	thirty thirty-seconds.
nine twentieths.	thirteen twenty-fourths.
four twenty-fifths.	fifteen nineteenths.

Read : $\frac{3}{6}$, $\frac{5}{13}$, $\frac{4}{9}$, $\frac{11}{21}$, $\frac{9}{22}$, $\frac{3}{17}$, $\frac{4}{19}$, $\frac{12}{23}$, $\frac{19}{25}$, $\frac{11}{27}$.

If the numerator is smaller than the denominator, the fraction is called a *proper fraction;* as $\frac{7}{8}$.

If the numerator is equal to the denominator, or greater than the denominator, the fraction is called an *improper fraction;* as $\frac{8}{8}$, $\frac{15}{8}$.

A *mixed number* is a whole number and a fraction; as $5\frac{2}{7}$, read five *and* two-sevenths.

Every whole number may be regarded as a fraction having 1 for the denominator.

Thus, 8 may be written $\frac{8}{1}$.

To change a whole number to an improper fraction:

We multiply the whole number by the denominator of the required fraction, and write the denominator under the product.

How many quarters of a dollar are there in $6 ?

Since there are four quarters ($\frac{4}{4}$) in one dollar, in 6 dollars there are 6 times four quarters ($\frac{4}{4}$), or twenty-four quarters ($\frac{24}{4}$).

Change to improper fractions :

$2 = \frac{}{2}$	$2 = \frac{}{3}$	$2 = \frac{}{4}$	$2 = \frac{}{5}$	$2 = \frac{}{6}$
$3 = \frac{}{2}$	$3 = \frac{}{3}$	$3 = \frac{}{4}$	$3 = \frac{}{5}$	$3 = \frac{}{6}$
$4 = \frac{}{2}$	$4 = \frac{}{3}$	$4 = \frac{}{4}$	$4 = \frac{}{5}$	$4 = \frac{}{6}$
$5 = \frac{}{2}$	$5 = \frac{}{3}$	$5 = \frac{}{4}$	$5 = \frac{}{5}$	$5 = \frac{}{6}$
$6 = \frac{}{2}$	$6 = \frac{}{3}$	$6 = \frac{}{4}$	$6 = \frac{}{5}$	$6 = \frac{}{6}$
$7 = \frac{}{2}$	$7 = \frac{}{3}$	$7 = \frac{}{4}$	$7 = \frac{}{5}$	$7 = \frac{}{6}$
$8 = \frac{}{2}$	$8 = \frac{}{3}$	$8 = \frac{}{4}$	$8 = \frac{}{5}$	$8 = \frac{}{6}$
$9 = \frac{}{2}$	$9 = \frac{}{3}$	$9 = \frac{}{4}$	$9 = \frac{}{5}$	$9 = \frac{}{6}$

$2 = \frac{}{8}$	$2 = \frac{}{9}$	$2 = \frac{}{10}$	$2 = \frac{}{12}$	$2 = \frac{}{16}$
$3 = \frac{}{8}$	$3 = \frac{}{9}$	$3 = \frac{}{10}$	$3 = \frac{}{12}$	$3 = \frac{}{16}$
$4 = \frac{}{8}$	$4 = \frac{}{9}$	$4 = \frac{}{10}$	$4 = \frac{}{12}$	$4 = \frac{}{16}$
$5 = \frac{}{8}$	$5 = \frac{}{9}$	$5 = \frac{}{10}$	$5 = \frac{}{12}$	$5 = \frac{}{16}$
$6 = \frac{}{8}$	$6 = \frac{}{9}$	$6 = \frac{}{10}$	$6 = \frac{}{12}$	$6 = \frac{}{16}$
$7 = \frac{}{8}$	$7 = \frac{}{9}$	$7 = \frac{}{10}$	$7 = \frac{}{12}$	$7 = \frac{}{16}$
$8 = \frac{}{8}$	$8 = \frac{}{9}$	$8 = \frac{}{10}$	$8 = \frac{}{12}$	$8 = \frac{}{16}$

Change :

1. 3 to 6ths.	**5.** 11 to 9ths.	**9.** 27 to 15ths.
2. 7 to 8ths.	**6.** 12 to 11ths.	**10.** 12 to 20ths.
3. 8 to 7ths.	**7.** 13 to 12ths.	**11.** 13 to 25ths.
4. 9 to 6ths.	**8.** 19 to 13ths.	**12.** 14 to 50ths.

To change a mixed number to an improper fraction:

We multiply the whole number by the denominator of the fraction, and to the product add the numerator; under this sum we write the denominator.

Change $6\frac{3}{4}$ to fourths.

Since there are 4 fourths ($\frac{4}{4}$) in 1, in 6 there are 6 times 4 fourths ($\frac{4}{4}$), or $\frac{24}{4}$; and 24 fourths ($\frac{24}{4}$) and 3 fourths ($\frac{3}{4}$) are 27 fourths ($\frac{27}{4}$).

Change to improper fractions:

$2\frac{1}{2}=\frac{}{2}$	$1\frac{1}{3}=\frac{}{3}$	$2\frac{5}{6}=\frac{}{6}$	$3\frac{1}{8}=\frac{}{8}$	$1\frac{1}{12}=\frac{}{12}$
$4\frac{1}{2}=\frac{}{2}$	$3\frac{1}{3}=\frac{}{3}$	$3\frac{1}{6}=\frac{}{6}$	$5\frac{7}{8}=\frac{}{8}$	$2\frac{7}{12}=\frac{}{12}$
$6\frac{1}{2}=\frac{}{2}$	$2\frac{1}{4}=\frac{}{4}$	$5\frac{1}{7}=\frac{}{7}$	$2\frac{1}{9}=\frac{}{9}$	$1\frac{5}{16}=\frac{}{16}$
$7\frac{1}{2}=\frac{}{2}$	$5\frac{1}{4}=\frac{}{4}$	$3\frac{5}{7}=\frac{}{7}$	$2\frac{5}{9}=\frac{}{9}$	$1\frac{11}{16}=\frac{}{16}$

Change to improper fractions:

1. $11\frac{1}{2}$.	11. $5\frac{31}{50}$.	21. $12\frac{1}{15}$.	31. $3\frac{22}{27}$.
2. $12\frac{3}{4}$.	12. $4\frac{31}{36}$.	22. $19\frac{1}{10}$.	32. $4\frac{15}{17}$.
3. $5\frac{11}{12}$.	13. $20\frac{3}{19}$.	23. $18\frac{7}{11}$.	33. $6\frac{13}{14}$.
4. $4\frac{13}{16}$.	14. $10\frac{5}{17}$.	24. $10\frac{25}{36}$.	34. $8\frac{11}{12}$.
5. $8\frac{4}{25}$.	15. $25\frac{4}{5}$.	25. $17\frac{7}{12}$.	35. $7\frac{5}{13}$.
6. $15\frac{5}{6}$.	16. $18\frac{2}{9}$.	26. $16\frac{6}{25}$.	36. $9\frac{7}{15}$.
7. $16\frac{6}{7}$.	17. $21\frac{3}{11}$.	27. $15\frac{9}{50}$.	37. $2\frac{47}{50}$.
8. $25\frac{7}{8}$.	18. $17\frac{6}{7}$.	28. $14\frac{9}{20}$.	38. $3\frac{37}{100}$.
9. $2\frac{87}{100}$.	19. $13\frac{4}{9}$.	29. $13\frac{13}{20}$.	39. $5\frac{91}{100}$.
10. $13\frac{3}{13}$.	20. $12\frac{5}{11}$.	30. $11\frac{21}{25}$.	40. $9\frac{49}{50}$.

To change an improper fraction to a whole or mixed number:

We divide the numerator by the denominator.

Thus, $\frac{25}{6} = 4\frac{1}{6}$.

The quotient will be a whole number or a mixed number. If a mixed number, the fractional part will have for numerator *the remainder of the division*, and for denominator *the divisor*.

Change to whole or mixed numbers the following:

1. $\frac{15}{5}$.
2. $\frac{18}{3}$.
3. $\frac{24}{12}$.
4. $\frac{54}{9}$.
5. $\frac{56}{8}$.
6. $\frac{63}{7}$.

7. $\frac{18}{5}$.
8. $\frac{29}{3}$.
9. $\frac{24}{7}$.
10. $\frac{54}{8}$.
11. $\frac{56}{9}$.
12. $\frac{63}{11}$.

13. $\frac{35}{2}$.
14. $\frac{23}{6}$.
15. $\frac{25}{8}$.
16. $\frac{53}{5}$.
17. $\frac{27}{4}$.
18. $\frac{40}{12}$.

19. $\frac{70}{16}$.
20. $\frac{80}{25}$.
21. $\frac{51}{14}$.
22. $\frac{37}{12}$.
23. $\frac{73}{36}$.
24. $\frac{91}{16}$.

To reduce a fraction to its lowest terms:

We divide out from both terms every number that will divide each term without a remainder.

Thus by dividing both terms of $\frac{8}{10}$ by 2 we get $\frac{4}{5}$.

Note. The Teacher must illustrate this example, and other examples until the pupils understand fully that the reduction of a fraction to lower terms does not alter its value.

Reduce to lowest terms:

1. $\frac{16}{24}$.
2. $\frac{14}{21}$.
3. $\frac{16}{20}$.
4. $\frac{25}{50}$.
5. $\frac{9}{21}$.
6. $\frac{2}{16}$.

7. $\frac{21}{27}$.
8. $\frac{9}{15}$.
9. $\frac{21}{28}$.
10. $\frac{28}{35}$.
11. $\frac{24}{30}$.
12. $\frac{25}{35}$.

13. $\frac{10}{45}$.
14. $\frac{25}{40}$.
15. $\frac{14}{42}$.
16. $\frac{25}{100}$.
17. $\frac{50}{100}$.
18. $\frac{75}{100}$.

19. $\frac{18}{27}$.
20. $\frac{33}{55}$.
21. $\frac{24}{32}$.
22. $\frac{50}{75}$.
23. $\frac{32}{36}$.
24. $\frac{21}{84}$.

MULTIPLICATION OF FRACTIONS.

If we take $\frac{1}{2}$ of $\frac{1}{4}$ of an apple, we have $\frac{1}{8}$ of an apple, and if we take $\frac{1}{2}$ of $\frac{3}{4}$ of an apple, we have $\frac{3}{8}$ of an apple.

NOTE. The Teacher should illustrate this by actually dividing an apple into quarters and then each quarter into halves.

That is, $\frac{1}{2}$ of $\frac{1}{4}=\frac{1}{8}$, and $\frac{1}{2}$ of $\frac{3}{4}=\frac{3}{8}$.　Hence,

To multiply one fraction by another :

We take the product of the numerators for the required numerator, and of the denominators for the denominator.

Mixed numbers and whole numbers may be written as improper fractions, and thus brought under the rule.

The work of multiplying fractions may be much shortened by **cancellation**; that is, by first dividing out every number that is contained in a numerator and a denominator without remainder.

Find the product of $\frac{6}{7}$, $2\frac{4}{5}$, and 3.

Now $2\frac{4}{5} = \frac{14}{5}$, and 3 may be written $\frac{3}{1}$.

Hence the product is $\dfrac{6 \times \overset{2}{\cancel{14}} \times 3}{\cancel{7} \times 5 \times 1} = \frac{36}{5} = 7\frac{1}{5}$.

Cancel the 7 from the denominator and from the 14 in the numerator, and then multiply ; we have $\frac{36}{5}$, or $7\frac{1}{5}$.

1. $\frac{1}{2}$ of $\frac{6}{7}=\frac{}{7}$.	7. $\frac{1}{4}$ of $\frac{8}{9}=\frac{}{9}$.	13. $\frac{2}{3}$ of $\frac{5}{6}=\frac{}{9}$.
2. $\frac{1}{2}$ of $\frac{8}{9}=\frac{}{9}$.	8. $\frac{1}{4}$ of $\frac{12}{13}=\frac{}{13}$.	14. $\frac{3}{8}$ of $\frac{4}{9}=\frac{}{6}$.
3. $\frac{1}{2}$ of $\frac{10}{11}=\frac{}{11}$.	9. $\frac{1}{4}$ of $\frac{16}{20}=\frac{}{20}$.	15. $\frac{2}{7}$ of $\frac{14}{3}=\frac{}{3}$.
4. $\frac{1}{3}$ of $\frac{6}{7}=\frac{}{7}$.	10. $\frac{1}{5}$ of $\frac{10}{11}=\frac{}{11}$.	16. $\frac{3}{8}$ of $\frac{8}{9}=\frac{}{3}$.
5. $\frac{1}{3}$ of $\frac{9}{11}=\frac{}{11}$.	11. $\frac{1}{5}$ of $\frac{15}{16}=\frac{}{16}$.	17. $\frac{11}{5}$ of $\frac{10}{33}=\frac{}{3}$.
6. $\frac{1}{3}$ of $\frac{15}{16}=\frac{}{16}$.	12. $\frac{1}{5}$ of $\frac{20}{21}=\frac{}{21}$.	18. $\frac{5}{8}$ of $\frac{24}{7}=\frac{}{7}$.

To multiply a mixed number by a whole number :

We multiply the fraction first, and then the integral part of the mixed number, and add the results.

Find the products of :

1. $2 \times 2\frac{1}{2}$.	8. $4 \times 2\frac{1}{3}$.	15. $5 \times 2\frac{1}{5}$.
2. $2 \times 3\frac{1}{2}$.	9. $4 \times 3\frac{1}{4}$.	16. $5 \times 2\frac{2}{5}$.
3. $2 \times 3\frac{1}{3}$.	10. $4 \times 2\frac{1}{8}$.	17. $5 \times 3\frac{3}{5}$.
4. $3 \times 3\frac{1}{3}$.	11. $4 \times 3\frac{1}{2}$.	18. $5 \times 4\frac{4}{5}$.
.5. $3 \times 2\frac{1}{6}$.	12. $4 \times 4\frac{1}{7}$.	19. $5 \times 1\frac{1}{8}$.
6. $3 \times 5\frac{1}{5}$.	13. $4 \times 4\frac{1}{5}$.	20. $5 \times 2\frac{1}{6}$.
7. $3 \times 3\frac{1}{6}$.	14. $4 \times 3\frac{2}{9}$.	21. $5 \times 3\frac{2}{10}$.

To multiply a whole number by a mixed number :

We multiply the whole number by the fraction first, and then by the integral part of the mixed number, and add the results.

Multiply 8 by $2\frac{1}{3}$.

$$\begin{array}{r} 8 \\ 2\frac{1}{3} \\ \hline 2\frac{2}{3} \\ 16 \\ \hline 18\frac{2}{3} \end{array}$$

Here we multiply 8 by $\frac{1}{3}$ and get $2\frac{2}{3}$.
Then we multiply 8 by 2 and get 16.
By adding the $2\frac{2}{3}$ and the 16 we obtain $18\frac{2}{3}$.

Find the products :

.1. $2\frac{1}{2} \times 6$.	6. $2\frac{1}{2} \times 12$.	11. $7\frac{1}{6} \times 21$.
2. $2\frac{1}{3} \times 6$.	7. $2\frac{3}{4} \times 8$.	12. $8\frac{3}{8} \times 22$.
3. $3\frac{1}{3} \times 6$.	8. $2\frac{2}{3} \times 9$.	13. $2\frac{1}{2} \times 6\frac{4}{5}$.
4. $3\frac{1}{2} \times 6$.	9. $3\frac{1}{2} \times 20$.	14. $3\frac{1}{3} \times 8\frac{1}{4}$.
5. $4\frac{1}{6} \times 6$.	10. $3\frac{5}{6} \times 12$.	15. $3\frac{1}{2} \times 6\frac{2}{7}$.

DIVISION OF FRACTIONS.

To divide $\frac{1}{2}$ of a dollar by $\frac{1}{4}$ of a dollar is to find the *number of quarters* of a dollar it is necessary to take in order to have half a dollar. It is obvious the number is 2. Hence,

$$\tfrac{1}{2} \div \tfrac{1}{4} = 2.$$
But $\tfrac{4}{1} \times \tfrac{1}{2} = 2.$

Therefore, to divide by $\frac{1}{4}$ gives the same result as to multiply by $\frac{4}{1}$.

Now $\frac{4}{1}$ is $\frac{1}{4}$ *inverted.* Therefore,

To divide by a fraction:

We invert the fraction and multiply.
Mixed numbers and whole numbers may be written as improper fractions, and thus brought under the rule.

Find the quotients:

1. $\frac{3}{2} \div \frac{3}{4}$.
2. $\frac{3}{4} \div \frac{3}{2}$.
3. $\frac{5}{9} \div \frac{2}{3}$.
4. $\frac{2}{3} \div \frac{5}{9}$.
5. $\frac{2}{3} \div \frac{3}{5}$.
6. $\frac{2}{3} \div \frac{5}{3}$.
7. $\frac{5}{3} \div \frac{2}{3}$.
8. $6 \div \frac{2}{5}$.
9. $\frac{2}{5} \div 6$.
10. $\frac{2}{7} \div \frac{3}{14}$.
11. $2 \div \frac{3}{5}$.
12. $\frac{2}{3} \div \frac{5}{6}$.

13. $2 \div 2\frac{1}{2}$.
14. $2 \div 3\frac{1}{2}$.
15. $3\frac{1}{3} \div 2$.
16. $3\frac{1}{2} \div 3$.
17. $2\frac{1}{4} \div 3$.
18. $5\frac{1}{4} \div 7$.
19. $2\frac{2}{5} \div 4$.
20. $4 \div 3\frac{1}{5}$.
21. $6 \div 1\frac{1}{2}$.
22. $4 \div 2\frac{1}{4}$.
23. $4 \div 1\frac{1}{3}$.
24. $8 \div 2\frac{2}{3}$.

25. $3\frac{1}{3} \div 2\frac{1}{2}$.
26. $4\frac{1}{2} \div 2\frac{1}{4}$.
27. $2\frac{1}{4} \div 4\frac{1}{2}$.
28. $2\frac{1}{2} \div 3\frac{1}{3}$.
29. $8\frac{1}{3} \div 4\frac{1}{6}$.
30. $4\frac{1}{6} \div 8\frac{1}{3}$.
31. $8\frac{1}{3} \div 1\frac{2}{3}$.
32. $5\frac{1}{4} \div 1\frac{3}{4}$.
33. $4\frac{2}{3} \div 3\frac{1}{2}$.
34. $5\frac{1}{3} \div 2\frac{2}{3}$.
35. $2\frac{2}{3} \div 5\frac{1}{3}$.
36. $7\frac{1}{7} \div 1\frac{3}{7}$.

SIMILAR FRACTIONS.

If both terms of a fraction are multiplied by the same number, the value of the fraction is not altered.

By this operation the *number* of parts is increased, and the *size* of the parts is decreased, at the same rate.

Fractions that have a common denominator are called *similar fractions*.

Reduce $\frac{1}{2}$, $\frac{2}{3}$, $\frac{3}{4}$ to similar fractions, having **12** for their common denominator.

We find the required numerators by dividing 12 by the denominator of the first fraction and multiplying the result by the numerator of the first fraction; and so proceed with each of the given fractions. Thus,

$12 \div 2 = 6$, and $1 \times 6 = 6$. Therefore $\frac{1}{2} = \frac{6}{12}$.
$12 \div 3 = 4$, and $2 \times 4 = 8$. Therefore $\frac{2}{3} = \frac{8}{12}$.
$12 \div 4 = 3$, and $3 \times 3 = 9$. Therefore $\frac{3}{4} = \frac{9}{12}$.

Hence the required fractions are $\frac{6}{12}$, $\frac{8}{12}$, $\frac{9}{12}$. Therefore,

To reduce fractions to similar fractions with a given common denominator:

We divide the given common denominator by the denominator of the first fraction, and multiply the quotient by its numerator, and this will be the required numerator of the first fraction. In the same way we find the numerator of each of the other fractions.

Reduce to similar fractions having for denominator the number given in parenthesis for each problem:

1. $\frac{1}{3}$, $\frac{3}{4}$, $\frac{5}{6}$ (12).　　6. $\frac{2}{3}$, $\frac{3}{8}$, $\frac{5}{24}$ (24).　　11. $\frac{5}{8}$, $\frac{5}{6}$, $\frac{5}{24}$ (24).

2. $\frac{1}{4}$, $\frac{1}{6}$, $\frac{1}{12}$ (12).　　7. $\frac{1}{2}$, $\frac{5}{7}$, $\frac{3}{14}$ (14).　　12. $\frac{3}{4}$, $\frac{11}{28}$, $\frac{2}{7}$ (28).

3. $\frac{1}{2}$, $\frac{2}{9}$, $\frac{5}{6}$ (18).　　8. $\frac{1}{3}$, $\frac{3}{7}$, $\frac{4}{21}$ (21).　　13. $\frac{5}{12}$, $\frac{7}{36}$, $\frac{5}{9}$ (36).

4. $\frac{1}{2}$, $\frac{3}{4}$, $\frac{5}{8}$ (8).　　9. $\frac{1}{5}$, $\frac{1}{3}$, $\frac{8}{15}$ (15).　　14. $\frac{2}{7}$, $\frac{3}{14}$, $\frac{5}{42}$ (42).

5. $\frac{1}{3}$, $\frac{5}{9}$, $\frac{7}{18}$ (18).　　10. $\frac{1}{3}$, $\frac{5}{7}$, $\frac{9}{14}$ (42).　　15. $\frac{3}{5}$, $\frac{7}{15}$, $\frac{8}{25}$ (75).

ADDITION OF FRACTIONS.

Add $\frac{1}{6}$, $\frac{1}{3}$, $\frac{3}{4}$.

These fractions changed to similar fractions with denominator 12 become $\frac{2}{12}$, $\frac{4}{12}$, $\frac{9}{12}$, and $\frac{2}{12} + \frac{4}{12} + \frac{9}{12} = \frac{15}{12}$.

But $\frac{15}{12} = \frac{5}{4} = 1\frac{1}{4}$.　Therefore,

To add fractions:

We change the fractions to similar fractions (if they are not similar), and write the sum of the numerators of the similar fractions over the common denominator.

We reduce the resulting fraction to its lowest terms; and if it is an improper fraction, we reduce it to a whole or mixed number.

Change to similar fractions and add:

$\frac{1}{2}+\frac{1}{4}=\frac{}{4}$	$\frac{1}{3}+\frac{1}{4}=\frac{}{12}$	$\frac{1}{4}+\frac{1}{6}=\frac{}{12}$	$\frac{1}{6}+\frac{1}{9}=\frac{}{18}$
$\frac{1}{2}+\frac{1}{8}=\frac{}{8}$	$\frac{1}{3}+\frac{1}{5}=\frac{}{15}$	$\frac{1}{4}+\frac{1}{8}=\frac{}{8}$	$\frac{1}{6}+\frac{1}{12}=\frac{}{12}$
$\frac{1}{2}+\frac{1}{3}=\frac{}{6}$	$\frac{1}{3}+\frac{1}{6}=\frac{}{6}$	$\frac{1}{4}+\frac{1}{10}=\frac{}{20}$	$\frac{1}{6}+\frac{1}{8}=\frac{}{24}$
$\frac{1}{2}+\frac{1}{6}=\frac{}{6}$	$\frac{1}{3}+\frac{1}{12}=\frac{}{12}$	$\frac{1}{4}+\frac{1}{12}=\frac{}{12}$	$\frac{1}{9}+\frac{1}{3}=\frac{}{9}$
$\frac{1}{2}+\frac{1}{5}=\frac{}{10}$	$\frac{1}{4}+\frac{1}{5}=\frac{}{20}$	$\frac{1}{4}+\frac{1}{16}=\frac{}{16}$	$\frac{1}{3}+\frac{1}{8}=\frac{}{24}$

$\frac{1}{2}=\frac{}{6}$	$\frac{1}{2}=\frac{}{8}$	$\frac{1}{2}=\frac{}{10}$	$\frac{1}{2}=\frac{}{12}$	$\frac{1}{2}=\frac{}{12}$
$\frac{1}{3}=\frac{}{6}$	$\frac{1}{4}=\frac{}{8}$	$\frac{1}{5}=\frac{}{10}$	$\frac{1}{6}=\frac{}{12}$	$\frac{1}{4}=\frac{}{12}$
$\frac{1}{6}=\frac{}{6}$	$\frac{1}{8}=\frac{}{8}$	$\frac{1}{10}=\frac{}{10}$	$\frac{1}{12}=\frac{}{12}$	$\frac{1}{12}=\frac{}{12}$

$\frac{1}{2}=\frac{}{16}$	$\frac{1}{4}=\frac{}{12}$	$\frac{1}{3}=\frac{}{12}$	$\frac{1}{3}=\frac{}{18}$	$\frac{1}{3}=\frac{}{12}$
$\frac{1}{8}=\frac{}{16}$	$\frac{1}{3}=\frac{}{12}$	$\frac{1}{6}=\frac{}{12}$	$\frac{1}{6}=\frac{}{18}$	$\frac{1}{4}=\frac{}{12}$
$\frac{1}{16}=\frac{}{16}$	$\frac{1}{12}=\frac{}{12}$	$\frac{1}{12}=\frac{}{12}$	$\frac{1}{9}=\frac{}{18}$	$\frac{1}{6}=\frac{}{12}$

$\frac{1}{2}=\frac{}{12}$	$\frac{1}{4}=\frac{}{12}$	$\frac{1}{2}=\frac{}{24}$	$\frac{1}{6}=\frac{}{24}$	$\frac{1}{6}=\frac{}{36}$
$\frac{1}{3}=\frac{}{12}$	$\frac{1}{3}=\frac{}{12}$	$\frac{1}{3}=\frac{}{24}$	$\frac{1}{8}=\frac{}{24}$	$\frac{1}{9}=\frac{}{36}$
$\frac{1}{6}=\frac{}{12}$	$\frac{1}{6}=\frac{}{12}$	$\frac{1}{8}=\frac{}{24}$	$\frac{1}{12}=\frac{}{24}$	$\frac{1}{4}=\frac{}{36}$

SUBTRACTION OF FRACTIONS.

Subtract $\frac{1}{6}$ from $\frac{3}{4}$.

These fractions changed to similar fractions having **12** for a denominator become $\frac{2}{12}$, $\frac{9}{12}$; and $\frac{9}{12} - \frac{2}{12} = \frac{7}{12}$. Hence,

To subtract one fraction from another :

We change the fractions to similar fractions (if they are not similar); then subtract the numerator of the subtrahend from that of the minuend, and write the remainder over the common denominator.

We reduce the resulting fraction to its lowest terms.

Change to similar fractions and subtract:

$\frac{1}{2} = \frac{}{4}$ $\frac{1}{2} = \frac{}{8}$ $\frac{1}{2} = \frac{}{6}$ $\frac{1}{2} = \frac{}{6}$ $\frac{1}{3} = \frac{}{6}$
$\frac{1}{4} = \frac{}{4}$ $\frac{1}{8} = \frac{}{8}$ $\frac{1}{6} = \frac{}{6}$ $\frac{1}{3} = \frac{}{6}$ $\frac{1}{6} = \frac{}{6}$

$\frac{1}{3} = \frac{}{9}$ $\frac{1}{3} = \frac{}{12}$ $\frac{1}{3} = \frac{}{15}$ $\frac{1}{3} = \frac{}{12}$ $\frac{3}{4} = \frac{}{4}$
$\frac{1}{9} = \frac{}{9}$ $\frac{1}{4} = \frac{}{12}$ $\frac{1}{5} = \frac{}{15}$ $\frac{1}{12} = \frac{}{12}$ $\frac{1}{2} = \frac{}{4}$

$\frac{2}{3} = \frac{}{6}$ $\frac{5}{8} = \frac{}{8}$ $\frac{3}{5} = \frac{}{10}$ $\frac{4}{5} = \frac{}{10}$ $\frac{4}{5} = \frac{}{20}$
$\frac{1}{2} = \frac{}{6}$ $\frac{1}{2} = \frac{}{8}$ $\frac{1}{2} = \frac{}{10}$ $\frac{1}{2} = \frac{}{10}$ $\frac{3}{4} = \frac{}{20}$

$\frac{3}{4} = \frac{}{20}$ $\frac{3}{5} = \frac{}{10}$ $\frac{1}{2} = \frac{}{10}$ $\frac{3}{5} = \frac{}{10}$ $\frac{7}{10} = \frac{}{10}$
$\frac{2}{5} = \frac{}{20}$ $\frac{3}{10} = \frac{}{10}$ $\frac{2}{5} = \frac{}{10}$ $\frac{3}{10} = \frac{}{10}$ $\frac{2}{5} = \frac{}{10}$

$\frac{7}{10} = \frac{}{10}$ $\frac{9}{10} = \frac{}{10}$ $\frac{15}{16} = \frac{}{16}$ $\frac{11}{16} = \frac{}{16}$ $\frac{3}{4} = \frac{}{12}$
$\frac{2}{5} = \frac{}{10}$ $\frac{4}{5} = \frac{}{10}$ $\frac{7}{8} = \frac{}{16}$ $\frac{1}{4} = \frac{}{16}$ $\frac{2}{3} = \frac{}{12}$

$\frac{1}{4} = \frac{}{16}$ $\frac{1}{2} = \frac{}{16}$ $\frac{3}{4} = \frac{}{12}$ $\frac{5}{6} = \frac{}{12}$ $\frac{3}{4} = \frac{}{12}$
$\frac{3}{16} = \frac{}{16}$ $\frac{7}{16} = \frac{}{16}$ $\frac{5}{12} = \frac{}{12}$ $\frac{3}{4} = \frac{}{12}$ $\frac{2}{3} = \frac{}{12}$

Change to similar fractions and add :

1. $\frac{2}{3}$ and $\frac{1}{4}$.

2. $\frac{3}{4}$ and $\frac{5}{6}$.

3. $\frac{5}{6}$ and $\frac{1}{4}$.

4. $\frac{2}{3}$ and $\frac{5}{6}$.

5. $\frac{5}{6}$ and $\frac{4}{9}$.

6. $\frac{2}{3}$ and $\frac{5}{9}$.

7. $\frac{2}{3}$ and $\frac{5}{8}$.

8. $\frac{7}{9}$ and $\frac{2}{3}$.

9. $\frac{4}{5}$ and $1\frac{3}{5}$.

10. $\frac{4}{5}$ and $\frac{3}{10}$.

11. $\frac{2}{7}$ and $\frac{5}{6}$.

12. $\frac{3}{4}$ and $\frac{5}{12}$.

13. $\frac{7}{6}$ and $\frac{7}{15}$.

14. $\frac{3}{8}$ and $\frac{5}{24}$.

15. $\frac{3}{4}$, $\frac{5}{6}$, and $\frac{1}{4}$.

16. $\frac{2}{3}$, $\frac{4}{9}$, and $\frac{5}{6}$.

17. $\frac{1}{12}$, $\frac{5}{8}$, and $\frac{2}{3}$.

18. $\frac{4}{5}$, $\frac{1}{10}$, and $\frac{1}{2}$.

19. $\frac{3}{5}$, $\frac{7}{10}$, and $\frac{2}{3}$.

20. $\frac{3}{4}$, $\frac{5}{7}$, and $\frac{1}{2}$.

21. $\frac{3}{4}$, $\frac{7}{9}$, and $1\frac{1}{7}$.

If any of the expressions are integers or mixed numbers :

We add together separately the fractions, and the integers, and then add the results. Thus,

Add $3\frac{3}{4}$, $2\frac{2}{3}$, and $1\frac{1}{12}$.

We first change the fractions to similar fractions,

$$\tfrac{3}{4} = \tfrac{9}{12}, \quad \tfrac{2}{3} = \tfrac{8}{12}, \quad \tfrac{1}{12} = \tfrac{1}{12}.$$

We then add these fractions,

$$\tfrac{9}{12} + \tfrac{8}{12} + \tfrac{1}{12} = \tfrac{18}{12} = \tfrac{3}{2} = 1\tfrac{1}{2}$$

then we add the integers, $3 + 2 + 1 = 6$

then we add the results and get $\qquad 7\frac{1}{2}$.

Find the sum of :

1. $3\frac{7}{9}$ and $4\frac{2}{3}$.

2. $5\frac{7}{8}$ and $3\frac{3}{4}$.

3. $4\frac{5}{8}$ and $8\frac{7}{16}$.

4. $6\frac{2}{3}$ and $7\frac{3}{4}$.

5. $7\frac{1}{2}$ and $9\frac{2}{3}$.

6. $5\frac{5}{6}$ and $9\frac{4}{5}$.

7. $8\frac{7}{20}$ and $5\frac{9}{10}$.

8. $3\frac{5}{8}$, $11\frac{1}{6}$, and $7\frac{7}{24}$.

9. $\frac{3}{4}$, $10\frac{2}{3}$, and $9\frac{1}{2}$.

10. $4\frac{3}{4}$, $8\frac{1}{2}$, and $5\frac{3}{8}$.

11. $1\frac{5}{16}$, $7\frac{7}{8}$, and $6\frac{3}{4}$.

12. $7\frac{1}{2}$, $8\frac{2}{3}$, and $7\frac{4}{5}$.

13. $5\frac{3}{8}$, $6\frac{7}{12}$, and $9\frac{2}{3}$.

14. $9\frac{5}{7}$, $3\frac{2}{3}$, and $8\frac{5}{6}$.

Change to similar fractions and subtract:

1. $\frac{3}{7}$ from $\frac{2}{3}$.

2. $\frac{3}{4}$ from $\frac{7}{8}$.

3. $\frac{1}{3}$ from $\frac{4}{5}$.

4. $\frac{3}{8}$ from $\frac{2}{3}$.

5. $\frac{5}{6}$ from $1\frac{1}{2}$.

6. $\frac{7}{16}$ from 1.

7. $\frac{17}{20}$ from 1.

8. $\frac{11}{15}$ from 1.

9. $\frac{7}{12}$ from 1.

10. $\frac{13}{17}$ from 1.

11. $\frac{29}{90}$ from $\frac{19}{30}$.

12. $\frac{3}{14}$ from $\frac{19}{56}$.

13. $\frac{3}{10}$ from $\frac{11}{20}$.

14. $\frac{3}{8}$ from $\frac{23}{24}$.

15. $\frac{3}{14}$ from $\frac{19}{42}$.

Adding the same number to both the subtrahend and the minuend does not alter their difference.

Thus, $8 - 6 = 2$; and if we add the same number, 7, to both 8 and 6 we have $15 - 13 = 2$. Hence,

To subtract a mixed number from a whole number, or from a mixed number:

We add such a fraction to the subtrahend as will make it a whole number, and add the same fraction to the minuend; then we subtract. Thus,

Subtract $3\frac{3}{4}$ from $7\frac{2}{3}$.

Here we add $\frac{1}{4}$ to the subtrahend $3\frac{3}{4}$ and obtain 4;
and we add $\frac{1}{4}$ to the minuend $7\frac{2}{3}$ and obtain $7\frac{11}{12}$.
Then we subtract 4 from $7\frac{11}{12}$ and obtain $3\frac{11}{12}$.

Subtract:

1. $1\frac{7}{8}$ from $13\frac{3}{8}$.

2. $2\frac{2}{3}$ from $10\frac{3}{4}$.

3. $3\frac{3}{4}$ from $11\frac{5}{6}$.

4. $4\frac{1}{2}$ from $12\frac{2}{3}$.

5. $6\frac{5}{6}$ from $20\frac{2}{3}$.

6. $5\frac{11}{12}$ from $19\frac{5}{6}$.

7. $8\frac{13}{15}$ from $60\frac{4}{5}$.

8. $9\frac{3}{7}$ from $51\frac{2}{21}$.

9. $13\frac{1}{3}$ from $15\frac{1}{2}$.

10. $13\frac{1}{2}$ from $15\frac{1}{3}$.

11. $27\frac{5}{6}$ from $70\frac{8}{9}$.

12. $27\frac{8}{9}$ from $70\frac{5}{6}$.

13. $29\frac{5}{6}$ from $69\frac{3}{4}$.

14. $29\frac{3}{4}$ from $69\frac{5}{6}$.

15. $20\frac{5}{16}$ from $40\frac{7}{8}$.

16. $20\frac{7}{8}$ from $40\frac{5}{16}$.

17. $\frac{5}{12}$ from 23.

18. $\frac{9}{10}$ from 26.

19. $6\frac{1}{4}$ from $8\frac{3}{8}$.

20. $5\frac{3}{10}$ from $7\frac{11}{20}$.

21. $3\frac{11}{12}$ from $5\frac{19}{24}$.

22. $6\frac{5}{8}$ from $7\frac{2}{3}$.

23. $7\frac{2}{7}$ from $9\frac{13}{21}$.

24. $5\frac{4}{7}$ from $8\frac{7}{9}$.

Simple Fractions. A fraction that has a whole number for its numerator and a whole number for its denominator is called a *simple fraction.*

Compound Fractions. A fraction of a fraction, of a mixed number, or of a whole number is called a *compound fraction.* Thus,

$\frac{1}{2}$ of $\frac{3}{4}$, $\frac{1}{2}$ of $2\frac{3}{4}$, and $\frac{1}{2}$ of 7 are compound fractions.

To change a compound fraction to a simple fraction:

We find the product of the numerators for the required numerator, and the product of the denominators for the required denominator.

NOTE 1. We first change the whole numbers and the mixed numbers to improper fractions.

NOTE 2. The method of **cancellation** should be used when possible.

Thus, change $\frac{4}{5}$ of $\frac{2}{11}$ of 22 to a simple fraction.

$$\frac{4}{5} \text{ of } \frac{2}{\cancel{11}} \text{ of } \frac{\cancel{22}^{\,2}}{1} = \frac{16}{5} = 3\frac{1}{5}.$$

Change to the simplest form:

1. $\frac{5}{8}$ of $\frac{2}{3}$ of 4.
2. $\frac{4}{5}$ of $\frac{5}{6}$ of $\frac{6}{7}$ of 8.
3. $\frac{3}{4}$ of $\frac{5}{6}$ of $\frac{4}{5}$ of 3.
4. $\frac{3}{7}$ of $\frac{36}{10}$ of $\frac{8}{27}$ of $2\frac{1}{4}$.

5. $\frac{9}{10}$ of $\frac{6}{7}$ of $3\frac{1}{2}$.
6. $\frac{1}{3}$ of $\frac{5}{6}$ of $1\frac{1}{2}$ of 4.
7. $\frac{3}{11}$ of $1\frac{2}{9}$ of $1\frac{1}{3}$ of $2\frac{1}{4}$.
8. $2\frac{1}{2}$ of $\frac{7}{9}$ of $2\frac{1}{4}$ of $\frac{13}{21}$.

Change to the simplest form and add:

9. $\frac{2}{3}$ of $\frac{3}{5}$ of $\frac{5}{8}$ and $\frac{2}{5}$ of $3\frac{2}{11}$ of $9\frac{1}{6}$.
10. $\frac{3}{4}$ of $\frac{5}{8}$ of $7\frac{1}{9}$ and $3\frac{1}{7}$ of $1\frac{5}{9}$ of $3\frac{3}{8}$.
11. $\frac{11}{14}$ of 9 of $6\frac{1}{8}$ and $\frac{4}{15}$ of $\frac{9}{21}$ of $\frac{5}{8}$ of 7.

Change to the simplest form and subtract:

12. $\frac{2}{3}$ of $\frac{3}{4}$ of $2\frac{1}{2}$ of $\frac{28}{45}$ from $\frac{2}{5}$ of $1\frac{7}{8}$ of $1\frac{1}{2}$ of $1\frac{5}{9}$.
13. $\frac{3}{7}$ of $\frac{2}{9}$ of $\frac{4}{5}$ of $8\frac{3}{4}$ from $\frac{5}{8}$ of $\frac{4}{5}$ of $1\frac{4}{7}$.
14. $1\frac{1}{3}$ of $1\frac{11}{45}$ of $7\frac{1}{2}$ from $\frac{4}{5}$ of $3\frac{1}{9}$ of 3 of $3\frac{3}{4}$.

Complex Fractions. A fraction that has a fraction in one or both of its terms is called a *complex fraction*.

Thus, $\dfrac{2}{3\frac{1}{4}}$, $\dfrac{2\frac{1}{2}}{3}$, and $\dfrac{2\frac{1}{4}}{3\frac{1}{4}}$ are complex fractions.

To change a complex fraction to a simple fraction:

We divide the numerator by the denominator.

NOTE. We first change the whole numbers and the mixed numbers to improper fractions, and compound fractions to simple fractions.

Change $\dfrac{8\frac{3}{4}}{12\frac{1}{2}}$ to a simple fraction :

$$8\tfrac{3}{4} = \tfrac{35}{4}, \text{ and } 12\tfrac{1}{2} = \tfrac{25}{2}.$$

$$\text{Then } \tfrac{35}{4} \div \tfrac{25}{2} = \frac{\overset{}{2}}{\underset{5}{\cancel{25}}} \times \frac{\overset{7}{\cancel{35}}}{\underset{2}{\cancel{4}}} = \tfrac{7}{10}.$$

Change to the simplest form :

1. $\dfrac{\frac{2}{7}}{\frac{3}{5}}$.

2. $\dfrac{\frac{2}{7}}{\frac{4}{7}}$.

3. $\dfrac{\frac{2}{7}}{\frac{3}{8}}$.

4. $\dfrac{\frac{5}{8}}{4}$.

5. $\dfrac{4}{\frac{3}{8}}$.

6. $\dfrac{2\frac{1}{3}}{1\frac{4}{15}}$.

7. $\dfrac{\frac{8}{9}}{20}$.

8. $\dfrac{6\frac{1}{2}}{3\frac{1}{4}}$.

9. $\dfrac{7\frac{1}{9}}{2\frac{2}{3}}$.

10. $\dfrac{3\frac{1}{2}}{2\frac{8}{9}}$.

11. $\dfrac{6\frac{1}{2}}{7\frac{1}{7}}$.

12. $\dfrac{6\frac{3}{7}}{8\frac{1}{4}}$.

13. $\dfrac{9\frac{3}{8}}{7\frac{1}{2}}$.

14. $\dfrac{8\cdot}{3\frac{4}{5}}$.

15. $\dfrac{36}{3\frac{3}{8} \text{ of } 6\frac{2}{3}}$.

16. $\dfrac{3\frac{3}{8} \text{ of } 4\frac{4}{9}}{\frac{2}{5} \text{ of } 6\frac{1}{4} \text{ of } \frac{3}{8}}$.

17. $\dfrac{1\frac{1}{5} \text{ of } 3\frac{1}{3}}{\frac{1}{4} \text{ of } 1\frac{1}{3} \text{ of } 1\frac{1}{4}}$.

18. $\dfrac{12\frac{2}{3} \text{ of } 1\frac{8}{19}}{1\frac{5}{9} \text{ of } 3\frac{3}{7}}$.

19. $\dfrac{8\frac{2}{5} \text{ of } 1\frac{4}{21}}{1\frac{13}{22} \text{ of } 2\frac{5}{14}}$.

20. $\dfrac{\frac{1}{4} \text{ of } 3\frac{3}{7} \text{ of } 2\frac{11}{12}}{\frac{1}{33} \text{ of } 8\frac{9}{14}}$.

21. $\dfrac{3\frac{1}{3} \text{ of } 2\frac{1}{10}}{\frac{1}{4} \text{ of } \frac{3}{8} \text{ of } \frac{2}{5}}$.

To find the fraction that one number is of another:

We take the number denoting the part for the numerator, and the number denoting the whole for the denominator.

Thus, suppose we wish to find the fraction that 6 is of 8.

Since 1 is $\frac{1}{8}$ of 8,

$$6 \text{ is } 6 \times \tfrac{1}{8} \text{ of } 8, \text{ or } \tfrac{6}{8} = \tfrac{3}{4}.$$

Here the number denoting the part is 6, and the number denoting the whole is 8.

Again, suppose we wish to find the fraction that $2\frac{2}{5}$ is of $7\frac{1}{2}$.

We first change the mixed numbers to improper fractions,

$$2\tfrac{2}{5} = \tfrac{12}{5}, \text{ and } 7\tfrac{1}{2} = \tfrac{15}{2}.$$

Then we change the improper fractions to similar fractions,

$$\tfrac{12}{5} = \tfrac{24}{10}, \text{ and } \tfrac{15}{2} = \tfrac{75}{10}.$$

The question becomes, what part of 75 is 24 ? and the answer is $\frac{24}{75}$.

What fraction of :

1. 12 is 6 ?	**11.** $3\frac{1}{2}$ is $\frac{1}{2}$?	**21.** 100 is $8\frac{1}{3}$?
2. 18 is 12 ?	**12.** $5\frac{1}{4}$ is 3 ?	**22.** 100 is $12\frac{1}{2}$?
3. 16 is 14 ?	**13.** $6\frac{1}{4}$ is 5 ?	**23.** 100 is $16\frac{2}{3}$?
4. 63 is 42 ?	**14.** 12 is $3\frac{1}{3}$?	**24.** 100 is $6\frac{1}{4}$?
5. 72 is 18 ?	**15.** 15 is $3\frac{3}{4}$?	**25.** 100 is $37\frac{1}{2}$?
6. 24 is 16 ?	**16.** $3\frac{3}{4}$ is 2 ?	**26.** 100 is $33\frac{1}{3}$?
7. 36 is 32 ?	**17.** $3\frac{3}{7}$ is 4 ?	**27.** 100 is $66\frac{2}{3}$?
8. 48 is 32 ?	**18.** 24 is $3\frac{3}{7}$?	**28.** 100 is $62\frac{1}{2}$?
9. 32 is 24 ?	**19.** 24 is $6\frac{2}{5}$?	**29.** 100 is $87\frac{1}{2}$?
10. 50 is 15 ?	**20.** 15 is $3\frac{1}{3}$?	**30.** 100 is $83\frac{1}{3}$?

To find the whole when a fractional part is given:

We divide the given part by the numerator of the fraction, and multiply the quotient by the denominator.   Thus,

Find the cost of a barrel of cranberries, when $\frac{3}{4}$ of a barrel costs $6.

If $\frac{3}{4}$ of a barrel costs $6, $\frac{1}{4}$ of a barrel will cost $\frac{1}{3}$ of $6, or $2.   If $\frac{1}{4}$ of a barrel costs $2, $\frac{4}{4}$ or a barrel will cost 4 × $2, or $8.

1. 20 is $\frac{5}{6}$ of what number?

2. 15 is $\frac{3}{7}$ of what number?

3. 22 is $1\frac{1}{2}$ of what number?

4. 18 is $\frac{3}{13}$ of what number?

5. 60 is $\frac{5}{4}$ of what number?

6. 80 is $\frac{5}{9}$ of what number?

7. $8\frac{2}{5}$ is $\frac{7}{8}$ of what number?

8. $9\frac{1}{5}$ is $\frac{7}{9}$ of what number?

9. $3\frac{1}{3}$ is $\frac{2}{7}$ of what number?

10. $11\frac{1}{5}$ is $\frac{7}{8}$ of what number?

11. If $\frac{3}{4}$ of an acre of land is worth $30, what is 1 acre worth?   What is the value of $2\frac{3}{4}$ acres?   Of $4\frac{5}{8}$ acres?

12. If $\frac{9}{10}$ of a ton of rye straw costs $27, what will $3\frac{1}{4}$ tons cost?

13. A man sold 63 acres of land, which was $\frac{7}{11}$ of all the land he owned.   How many acres had he left?

14. If $\frac{7}{8}$ of a barrel of sugar can be bought for $8\frac{3}{4}$, how many barrels can be bought for $62\frac{1}{2}$?

15. If $\frac{3}{4}$ of a bushel of wheat is worth 54 cents, how many bushels can be bought for $36?

16. If a freight train goes 20 miles an hour, and goes only $\frac{5}{11}$ as fast as an express train, how far will the express train go in $3\frac{1}{4}$ hours?

CONVERSION OF FRACTIONS.

A decimal fraction is a common fraction whose denominator is one of the numbers 10, 100, 1000, etc.

Thus, 0.4 is the same as $\frac{4}{10}$.

To convert a decimal fraction to a common fraction :

We take for the numerator the entire number obtained after removing the decimal point, and for the denominator, 1 followed by as many zeros as there are decimal places in the original fraction; and reduce the resulting fraction to its lowest terms.

Thus, $3.25 = \frac{325}{100} = \frac{13}{4} = 3\frac{1}{4}$.

To convert a common fraction to a decimal fraction :

We divide the numerator by the denominator.

Thus,
$$\frac{1}{8} = \frac{1.000}{8} = 0.125.$$
$$\frac{4}{7} = \frac{4.000}{7} = 0.571\frac{3}{7}.$$
$$\frac{2}{3} = \frac{2.00}{3} = 0.666\frac{2}{3}.$$

Note. If the division does not terminate at the third decimal place, three decimal places will be sufficiently accurate for most problems. If the number at the *fourth* decimal place is greater than 5, we add 1 to the third decimal figure; if it is equal to 5, we carry the decimal to four places. Thus, $\frac{4}{7} = 0.571$, $\frac{2}{3} = 0.667$, and $\frac{7}{16} = 0.4375$.

Change to common fractions :

1. 0.08.	4. 0.375.	7. 0.425.	10. 3.125.
2. 0.625.	5. 0.004.	8. 0.015.	11. 1.725.
3. 0.032.	6. 0.256.	9. 7.075.	12. 7.875.

Change to decimal fractions :

13. $\frac{3}{50}$.	16. $\frac{1}{25}$.	19. $\frac{1}{250}$.	22. $7\frac{3}{40}$.
14. $\frac{3}{20}$.	17. $\frac{27}{200}$.	20. $17\frac{7}{8}$.	23. $1\frac{15}{16}$.
15. $\frac{1}{40}$.	18. $\frac{4}{125}$.	21. $5\frac{3}{8}$.	24. $5\frac{1}{16}$.

ORAL EXERCISES.

1. How many inches in $\frac{3}{4}$ of a yard?

2. How many ounces in $\frac{5}{8}$ of a pound?

3. How many pounds in $\frac{1}{2}$ of a ton?

4. How many cubic feet in $\frac{8}{9}$ of a cubic yard?

5. How many square rods in $\frac{3}{4}$ of an acre?

6. How many cord feet in $\frac{7}{8}$ of a cord?

7. How many pints in $\frac{3}{16}$ of a gallon?

8. How many hours in $\frac{2}{3}$ of a day?

9. How many minutes in $\frac{3}{4}$ of an hour?

10. How many quarters of a pound in 2 pounds?

11. How many quarters of a dollar in $\$6\frac{3}{4}$? in $\$7\frac{1}{4}$?

12. How many halves of an apple in $4\frac{1}{2}$ apples?

13. I have a string $2\frac{3}{4}$ yards long. Into how many pieces, each $\frac{1}{4}$ yard long, can I cut it?

14. How many gallons will 10 bottles hold if each bottle holds $\frac{1}{5}$ of a gallon?

15. Find the price of $2\frac{1}{4}$ dozen of eggs at 16 cents a dozen.

16. Find the price of $3\frac{1}{2}$ pounds of sugar at 6 cents a pound.

17. How many miles will a man walk in 2 hours at the rate of $3\frac{1}{2}$ miles an hour?

18. How many miles will a man walk in $2\frac{1}{2}$ hours at the rate of 3 miles an hour?

19. How many miles will a man walk in 3 hours at the rate of $3\frac{1}{2}$ miles an hour?

20. Express 2 ft. 6 in. as the fraction of a yard.

21. At $\$7$ a ton what is the cost of $\frac{3}{4}$ of a ton of coal?

22. At $\$6$ a cord what is the cost of $\frac{7}{8}$ of a cord of wood?

23. At 80 cents a bushel what is the cost of $2\frac{1}{4}$ bushels of Baldwin apples?

SLATE EXERCISES.

Find the cost, reckoning every fraction of a cent as a cent :

1. 3¾ doz. of eggs at 24 cents a dozen.
2. 3½ lb. of steak at 23 cents a pound.
3. 2½ lb. of tea at 65 cents a pound.
4. 17¾ yd. of muslin at 10 cents a yard.
5. 50 cans of tomatoes at $1.25 a dozen.
6. 2¾ bu. of potatoes at 18 cents a peck.
7. 16 bu. of oats at 37½ cents a bushel.
8. 24 bags of corn at $1.12½ a bag.
9. 36 bu. of wheat at 87½ cents a bushel.
10. 4 lb. and 12 oz. of butter at 20 cents a pound.
11. 8 lb. and 10 oz. of mutton at 12 cents a pound.
12. 6 qt. of molasses at 56 cents a gallon.
13. 43½ yd. of cotton cloth at 7 cents a yard.
14. 14 lb. 6 oz. of ham at 14 cents a pound.
15. 6 bu. and 3 pk. of wheat at 92 cents a bushel.
16. 2680 lb. of hay at $22 a ton.
17. 2 t. 8 cwt. of coal at $5.60 a ton.
18. 3½ bbl. of flour at $5.50 a barrel.
19. 2¾ bu. of cranberries at 7 cents a quart.
20. 3 pk. and 4 qt. of cranberries at $2.56 a bushel.
21. 2 cd. and 6 cu. ft. of wood at $3.50 a cord.
22. A pile of wood 26 ft. long, 4 ft. wide, and 5 ft. high at $3.84 a cord.
23. 4 doz. and 8 eggs at 30 cents a dozen.
24. 75,000 bricks at $6.75 a thousand.
25. 9 shares of stock at $98¼ a share.

REVIEW.

1. A merchant sold $\frac{1}{3}$ of a piece of cloth and afterwards $\frac{1}{2}$ of the remainder. What part of the piece had he left?

2. What part of a foot must be added to the sum of $\frac{1}{4}$ and $\frac{1}{3}$ of a foot to make one foot?

3. At $\frac{3}{4}$ of a dollar a yard, what is the cost of 6 yards of cloth?

4. At $5.50 a ton, what is the cost of $\frac{1}{2}$ of a ton of coal?

5. Three packages of sugar weigh respectively $2\frac{3}{16}$, $3\frac{7}{16}$, and $4\frac{5}{8}$ pounds. What is the weight of the whole?

6. If $2\frac{1}{4}$ pecks are sold from a bushel of potatoes, how many pecks remain?

7. Change to improper fractions: $6\frac{1}{3}$, $5\frac{2}{3}$, $6\frac{5}{9}$, $8\frac{1}{4}$, $12\frac{2}{3}$.

8. Change to mixed numbers: $\frac{19}{2}$, $\frac{25}{7}$, $\frac{29}{8}$, $\frac{31}{11}$, $\frac{41}{12}$, $\frac{98}{9}$.

9. Reduce to lowest terms: $\frac{16}{20}$, $\frac{9}{12}$, $\frac{56}{72}$, $\frac{42}{63}$, $\frac{72}{108}$.

10. At $\frac{2}{3}$ of a dollar a bushel, how many bushels of apples can be bought for 10 dollars?

11. If $\frac{7}{8}$ of a yard of cloth costs 42 cents, what will $2\frac{1}{4}$ yards of the cloth cost?

12. If $\frac{4}{5}$ of a cord of wood is sold for $4, what will $5\frac{5}{8}$ cords bring?

13. A farmer has $\frac{4}{5}$ of his cows in his large barn and the rest of them in a small barn. If he has 20 cows in the large barn, how many has he in the small barn?

14. If a barrel holds $2\frac{3}{4}$ bushels of apples, how many barrels will be needed for 264 bushels?

15. A man bought 25 pounds of sugar at the rate of 20 pounds for $1; and gave a two-dollar bill in payment. How much change is due him?

REVIEW.

1. Four fifths of $20 is $\frac{2}{7}$ of how much money?

2. What part of a ton of coal at $6 a ton will $\frac{3}{4}$ of a cord of wood pay for, if the wood is worth $4 a cord?

3. If $9\frac{3}{4}$ pounds of butter are sold from a firkin containing 15 pounds, how many pounds will remain in the firkin?

4. If $\frac{2}{3}$ of a yard of silk costs $\frac{4}{5}$, how many yards can be bought for $8\frac{2}{3}$?

5. If $\frac{2}{3}$ of the distance between two towns is $4\frac{1}{5}$ miles, how many miles are the two towns apart?

6. A carpet dealer sold $\frac{3}{4}$ of $\frac{2}{3}$ of a roll of carpet. What part of the roll had he left?

7. It takes $2\frac{3}{4}$ bushels of apples to make a barrel. How many barrels will $24\frac{3}{4}$ bushels make?

8. If $2\frac{1}{2}$ quarts of vinegar have been used from a gallon, what part of the gallon remains?

9. A boy after selling $\frac{1}{2}$ and $\frac{1}{3}$ of his marbles had 12 marbles left. How many had he at first?

10. A man sold a horse for $200 and lost $\frac{1}{5}$ of what the horse cost him. How much did he pay for the horse?

11. A man sold a horse for $180 and gained $\frac{1}{5}$ of what the horse cost him. How much did he pay for the horse?

12. If 54 yards are cut from a piece of cloth containing 81 yards, what part of the piece will remain?

13. If $7\frac{1}{2}$ barrels of flour cost $30, what will 91 barrels cost at the same rate?

14. Captain Newman owned $\frac{3}{4}$ of a ship. He sold $\frac{2}{3}$ of his share. What part of the ship did he sell? What part of the ship did he still own?

REVIEW.

1. Subtract $19\frac{6}{7}$ from 200.

2. Reduce to its simplest form $\frac{2}{3}$ of $\frac{7}{8}$ of $1\frac{5}{7}$.

3. Add $23\frac{1}{2}$, $17\frac{2}{3}$, $16\frac{3}{4}$, $11\frac{5}{12}$.

4. Divide $9\frac{1}{6}$ by $5\frac{1}{2}$.

5. A boy has $\frac{1}{2}$ of a bushel of walnuts. If he sells $\frac{3}{4}$ of them, what part of a bushel will he have left?

6. Find the value of $\frac{2}{3}$ of a chest of tea containing $58\frac{1}{2}$ pounds at $62\frac{1}{2}$ cents a pound.

7. If a man works $8\frac{1}{3}$ hours a day he can finish a certain piece of work in 12 days. How many days will it take him if he works 10 hours a day?

8. If $\frac{2}{7}$ of $\frac{2}{3}$ of a piece of land is worth $400, what is the whole piece worth?

9. Reduce $\frac{3}{5}$, $\frac{1}{4}$, and $\frac{7}{8}$ to decimal fractions and find the sum.

10. If $\frac{3}{4}$ of a yard of cloth costs $\frac{7}{8}$ of a dollar, what will $6\frac{2}{3}$ yards cost?

11. A farmer sold to a merchant 16 bushels of potatoes at $\frac{3}{4}$ of a dollar a bushel and bought 7 yards of cloth at $$1\frac{5}{6}$ per yard. Find the balance due.

12. If $\frac{2}{3}$ of $\frac{1}{5}$ of a yard of velvet is worth $$1\frac{3}{7}$, what is the value of 7 yards?

13. At $6\frac{1}{4}$ cents a yard, how many yards of cotton cloth can be bought for $2.25?

14. One third of a pole is blue, $\frac{2}{7}$ of it is red, and the rest is white. What part of the pole is white?

15. At $22\frac{1}{2}$ cents a yard, how many yards of sheeting can be bought for $3.15?

16. Multiply the sum of $\frac{9}{10}$ and $\frac{3}{5}$ by their difference.

REVIEW.

1. If a man steps on the average $2\frac{3}{4}$ feet, how many steps will he take in walking a mile (5280 ft.) ?

2. If $5\frac{1}{2}$ acres of land cost $550, what will $32\frac{1}{2}$ acres cost ?

3. If $4\frac{3}{4}$ tons of coal cost $28\frac{1}{2}$, what is the cost of $7\frac{1}{4}$ tons ?

4. If 5 pecks of potatoes cost $1.50, what will $6\frac{3}{4}$ bushels cost at the same rate ?

5. If $\frac{1}{3}$ of a certain farm is pasture, $\frac{3}{8}$ of it is in crops, and the remainder, 56 acres, is woodland, of how many acres does the farm consist ?

6. If it is $11\frac{1}{4}$ feet round a wagon wheel, how many times must the wheel turn in going a mile ?

7. What must be paid for a pile of wood 24 feet long, 5 feet high, and 4 feet wide, at $3\frac{1}{2}$ a cord ?

8. If $5\frac{1}{4}$ pounds of candy are divided equally among 7 boys, what part of a pound will each boy receive ?

9. From a cask containing $31\frac{1}{2}$ gallons of oil, 3 quarts a day are drawn out for 21 days. How many gallons are left in the cask ?

10. Subtract the sum of $\frac{2}{3}$, $\frac{3}{4}$, $\frac{5}{6}$, $\frac{8}{9}$, and $1\frac{1}{2}$ from 7.

11. At the rate of $3\frac{1}{4}$ miles an hour a man walks a certain distance in $3\frac{1}{2}$ hours. How many hours will it take him to walk the same distance at the rate of $3\frac{1}{8}$ miles an hour ?

12. The circumference of a circle is $3\frac{5}{113}$ of the diameter. Find the circumference of a circle if the diameter is $4\frac{1}{3}$ feet.

PERCENTAGE.

A **percentage** of a number is the result obtained by taking a stated number of **hundredths** of it.

One **hundredth** of a number is called one **per cent** of it; two **hundredths**, two **per cent**; three **hundredths**, three **per cent**; and so on.

This sign % stands for the words **per cent**.

Thus, 5 % of 300 means 0.05 of 300.
15½ % of 300 means 0.15½ of 300.
¼ % of 300 means 0.00¼ of 300.

When the per cent can be expressed as a common fraction in *small terms*, it is better to write it as a common fraction.

50 % of a number is $\frac{50}{100}$ or $\frac{1}{2}$ the number.

25 % of a number is $\frac{25}{100}$ or $\frac{1}{4}$ the number.

75 % of a number is $\frac{75}{100}$ or $\frac{3}{4}$ the number.

12½ % of a number is $\frac{12\frac{1}{2}}{100}$ or $\frac{1}{8}$ the number.

8⅓ % of a number is $\frac{8\frac{1}{3}}{100}$ or $\frac{1}{12}$ the number.

16⅔ % of a number is $\frac{16\frac{2}{3}}{100}$ or $\frac{1}{6}$ the number.

33⅓ % of a number is $\frac{33\frac{1}{3}}{100}$ or $\frac{1}{3}$ the number.

66⅔ % of a number is $\frac{66\frac{2}{3}}{100}$ or $\frac{2}{3}$ the number.

20 % of a number is $\frac{20}{100}$ or $\frac{1}{5}$ the number.

125 % of a number is $\frac{125}{100}$ or $\frac{5}{4}$ the number.

Express as per cent:

$\frac{1}{2}$ $\frac{1}{4}$ $\frac{1}{6}$ $\frac{2}{3}$ $\frac{5}{6}$ $\frac{3}{8}$ $\frac{1}{10}$ $\frac{7}{50}$

$\frac{1}{8}$ $\frac{1}{5}$ $\frac{3}{4}$ $\frac{4}{5}$ $\frac{1}{8}$ $\frac{5}{8}$ $\frac{1}{12}$ $\frac{3}{25}$

Find $16\frac{1}{3}\%$ of 336.

336	$16\frac{1}{3}\% = 0.16\frac{1}{3}$.
0.16$\frac{1}{3}$	

112
2016
336

54.88   **54.88.** *Ans.*

Find $16\frac{2}{3}\%$ of 336.

$16\frac{2}{3}\% = \frac{1}{6}$.

$\frac{1}{6}$ of 336 = 56.

56. *Ans.*

Hence,

To find a per cent of a number :

We multiply the number by the given per cent.

Find :

1. 6% of 175.
2. 25% of 300.
3. $16\frac{2}{3}\%$ of 480 men.
4. $5\frac{1}{3}\%$ of 675 sheep.
5. 10% of 1560 days.

6. $33\frac{1}{3}\%$ of $840.
7. 50% of 1216 oz.
8. $66\frac{2}{3}\%$ of 1518 lb.
9. 75% of 2040 ft.
10. $12\frac{1}{2}\%$ of 1648 mi.

To find the per cent one given number is of another :

We divide the number which represents the percentage by the other number; carrying the division to hundredths.

What per cent of 9 is 3 ?

Since 1 is $\frac{1}{9}$ of 9, 3 is $3 \times \frac{1}{9} = \frac{3}{9} = \frac{1}{3}$; and $\frac{1}{3} = 0.33\frac{1}{3} = 33\frac{1}{3}\%$.

The same result is obtained if we divide 3 by 9.

$$9\overline{)3.00}$$
$$0.33\frac{1}{3} = 33\frac{1}{3}\%.$$

What per cent of

1. 90 is 30 ?
2. 960 is 24 ?
3. 30 is 90 ?
4. 24 is 960 ?
5. 4108.5 is 821.7 ?

6. 2740 mi. are 548 mi. ?
7. 36 in. are 27 in. ?
8. $2.75 are $0.35 ?
9. 2240 lb. are 2000 lb. ?
10. 7000 gr. are 5760 gr. ?

To find a number when a percentage of the number and the rate per cent are given:

We express the rate per cent as a fraction; then we divide the percentage by the numerator of this fraction and multiply the quotient by the denominator.

If 400 is $12\frac{1}{2}\%$ of a number, what is the number?

Since 400 is $12\frac{1}{2}\%$, or $\frac{1}{8}$, of the number, the number is 8×400, or 3200.

1800 is $12\frac{1}{2}\%$ more than what number?

Since	100%	of the number = the number,
and	$12\frac{1}{2}\%$	of the number = the increase,

$112\frac{1}{2}\%$, or $\frac{9}{8}$, of the number = 1800.

Therefore, $\frac{1}{8}$ of the number $= \frac{1}{9}$ of 1800, or 200,

and $\frac{8}{8}$, or the number $= 8 \times 200$, or 1600.

1400 is $12\frac{1}{2}\%$ less than what number?

Since	100%	of the number = the number,
and	$12\frac{1}{2}\%$	of the number = the decrease,

$87\frac{1}{2}\%$, or $\frac{7}{8}$, of the number = 1400.

Therefore, $\frac{1}{8}$ of the number $= \frac{1}{7}$ of 1400, or 200,

and $\frac{8}{8}$, or the number $= 8 \times 200$, or 1600.

1. 21 is 75% of what number?
2. 25 is $62\frac{1}{2}\%$ of what number?
3. 20 is $16\frac{2}{3}\%$ of what number?
4. 30 is $37\frac{1}{2}\%$ of what number?
5. 60 is $66\frac{2}{3}\%$ of what number?
6. 18 is $33\frac{1}{3}\%$ of what number?
7. 21 is $87\frac{1}{2}\%$ of what number?
8. 700 is $16\frac{2}{3}\%$ more than what number?
9. 600 is 25% less than what number?
10. 240 is 20% less than what number?
11. 240 is 20% more than what number?
12. 360 is 10% less than what number?
13. 660 is 10% more than what number?
14. 170 is $6\frac{1}{4}\%$ more than what number?

1. A man sold a farm for $2000 and gained 25% on the sum he paid for it. How much did he pay for it?

2. A farmer had 40 lambs and sold 62½% of them. How many lambs had he left?

3. A fruit dealer buys oranges at $2 a hundred and sells them at a gain of 20%. How much does he get for them a hundred?

4. A man sold a house for $240 less than it cost him and lost 12% of its cost. How much did the house cost him?

5. If I buy potatoes at 50 cents a bushel and sell them for 55 cents a bushel, what per cent do I gain?

6. If I sell a horse for $150 and gain 25% of the cost, how much did I pay for the horse?

7. If I sell a horse for $150 and lose 25% of the cost, how much did I pay for the horse?

8. A house worth $5000 rents for $400 a year. What per cent of its value is the rent?

9. How many pounds of cheese bought at 10 cents a pound must be sold at a gain of 20% in order to make a net profit of $10?

10. If copper matte yields 55% of pure copper, how much copper will be obtained from a long ton (2240 pounds) of matte?

11. A man spent 55% of his money and had $65.70 left. How much had he at first?

12. A man sold 2000 bushels of wheat at a gain of 12½% of the cost. If his net gain is $200, how much did he pay for the wheat per bushel?

1. An agent buys goods to the amount of $5000 and charges a commission of 2%. How much is his commission ?

NOTE. Money paid to an agent for his services in buying or selling goods, in collecting debts, etc., is called **commission**.

2. An agent in Minneapolis buys 1000 barrels of flour at $4 a barrel. How much is his commission at 2½% ?

3. If an agent in New Orleans buys cotton for $8000 and charges 1½% commission, how much money must be sent to him to pay for the cost of the cotton and for his commission ?

4. An agent received $200 for selling goods. If his commission was 5%, for how much did he sell the goods ?

5. An agent sold goods for $1500 and received for his services $30. What rate of commission did he charge ?

6. An agent sold a farm for $16,000 and charged 5% commission. How much did the owner of the farm receive ?

7. An agent sold 450 bushels of wheat at 80 cents a bushel and received $9 for his commission. What rate of commission did he charge ?

8. An agent received $150 as his commission at 3% for buying 1200 barrels of flour. What price per barrel did he pay for the flour ?

9. At an average price of 50 cents a bushel and a commission of 4% for buying, how many bushels of potatoes can be bought for $78 ?

10. Find the total cost of 4000 barrels of flour delivered in Boston, if an agent in Chicago pays $3 a barrel for it, charges 3% commission, and pays 40 cents a barrel for freight from Chicago to Boston.

1. Find the premium for an insurance policy of $8000 at $1\frac{1}{2}\%$.

NOTE. In insurance, the money paid for a guaranty of a stated sum, in the event of loss by fire or other causes, is called the **premium**; the guaranty or written contract is called the **policy**.

2. Find the premium for an insurance policy of $6000 on a dwelling house at $1\frac{1}{2}\%$.

3. Find the premium for an insurance policy on a stock of goods of $9000 at 2%.

4. How much is the insurance on a store, if the cost of the insurance is $130 and the rate of the insurance is 1% ?

5. How much is the insurance on a house, if the cost of the insurance is $60 and the rate of the insurance is $1\frac{1}{4}\%$?

6. A store costing $9000 is insured for $\frac{2}{3}$ of its value at 1% premium. What is the owner's loss, including premium paid, if the store is destroyed by fire ?

7. A store costing $12,000 is insured for $\frac{2}{3}$ of its value at $1\frac{1}{4}\%$ premium. What is the insurance company's net loss if the store is destroyed by fire ?

8. What is the cost of insuring 6000 bushels of wheat at $1\frac{1}{4}\%$, if the wheat is reckoned at 90 cents a bushel ?

9. An insurance company charges $30 for insuring a house for $6000. What is the rate of insurance ?

10. An insurance company charges $48 for insuring a house for $6400. What is the rate of insurance ?

11. A building valued at $25,000 is insured in three companies. The first company takes $\frac{1}{4}$ of the risk; the second takes $\frac{2}{5}$ of the risk; and the third takes the remainder. If the building is damaged by fire to the amount of $2000, what must each company pay ?

1. A man sold furniture to the amount of $2500 and took off 5% from the bill for cash. How much was the discount?

Note. The sum taken off the list price of an article, from the amount of a bill or from a debt, is called discount.

2. Find the discount of 10% from a bill of $410.

3. Find the discount of 15% from a bill of $1600.

4. Find the discount of 12½% from a bill of $200.

5. Find the discount of 16⅔% from a bill of $300.

6. Find the discount of 25% from a bill of $50.

7. How much must be paid for a book marked $2, if 25% is taken off the marked price?

8. How much must be paid for a sewing machine, if the marked price is $30 and the discount is 33⅓%?

9. How much must be paid for a bicycle, if the marked price is $75 and the discount is 40%?

10. How much must be paid for a desk, if the marked price is $40 and the discount is 15%?

11. How much must be paid for a bookcase, if the marked price is $45 and the discount is 20%?

12. Find the net amount of a bill of $1300, if a discount of 5% is allowed for cash.

13. If the amount of a bill for books is $82 and a discount of 25% is allowed, how much will settle the bill?

14. How much money will be required to pay a debt of $1600, if a discount of 37½% is allowed?

15. A merchant bought cloth at $3.20 a yard. He sold it at a profit of 12½%, and took off from the bill 5% for cash. How many cents a yard did he make on the cloth?

INTEREST.

Money paid for the use of money is called **Interest**.

The money at interest is called the **Principal**.

The sum of the interest and principal is called the **Amount**.

To find the interest for a given number of months at 6%:

We put the decimal point two places to the left in the principal, and multiply by one-half the number of months.

Find the interest on $630 for 4 mo. at 6%.

$$\begin{array}{r} \$6.30 \\ 2 \\ \hline \$12.60 \end{array}$$

Here we put the decimal point *two places to the left* in the principal, and multiply by 2 ; that is, by ½ of 4.

To find the interest for any other rate than 6%:

We find the interest at 6%, divide the result by 6, and multiply the quotient by the given rate.

Find the interest on $630 for 4 mo. at 4½%.

The interest at 6% is $12.60.
The interest at 1% is ⅙ of $12.60, or $2.10.
The interest at 4½% is 4½ × $2.10, or $9.45.

Find the interest on :

1. $1220.40 for 3 mo. at 6%.
2. $2512.80 for 4 mo. at 5%.
3. $2084.20 for 1 mo. at 4½%.
4. $4500.60 for 5 mo. at 5½%.
5. $7508.50 for 6 mo. at 3½%.
6. $8501.20 for 3 mo. at 5%.
7. $9056.75 for 7 mo. at 6%.
8. $1000 for 4 mo. at 6%.
9. $1500 for 6 mo. at 4%.
10. $75.50 for 4 mo. at 5%.

To find the interest for a given number of days at 6%:

We put the decimal point three places to the left in the principal, and multiply by one-sixth of the number of days.

Find the interest on $7260 for 90 days at 6%.

$7.260 | Here we put the decimal point *three* places to the
  15 | left in the principal, and multiply by 15; that is, by ⅙
$108.900 | of 90.

To find the interest for any other rate than 6%:

We find the interest at 6%, divide the result by 6, and multiply the quotient by the given rate.

Find the interest on :

   1. $3600 for 30 days at 6%.
   2. $4500 for 33 days at 6%.
   3. $8000 for 93 days at 6%.
   4. $9875 for 60 days at 5%.
   5. $2525 for 63 days at 4½%.
   6. $3750 for 90 days at 3½%.
   7. $15.80 for 63 days at 4%.
   8. $256.40 for 45 days at 5½%.
   9. $645.25 for 123 days at 3%.

Find the amount (interest + principal) of :

   10. $750.25 for 123 days at 6%.
   11. $660.40 for 120 days at 6%.
   12. $525.80 for 93 days at 5%.
   13. $551.75 for 90 days at 4½%.
   14. $612.25 for 60 days at 3½%.
   15. $809.18 for 63 days at 5½%.
   16. $729.20 for 33 days at 4%.
   17. $505.90 for 30 days at 4½%.
   18. $819.78 for 33 days at 4¼%.

To find the interest for a given number of years:

We multiply the principal by the given rate per cent, and this product by the number of years.

Find the interest on $630 for 4 years at 5%.

> The interest for 1 year at 5% is 0.05 of $630, or $31.50.
> The interest for 4 years at 5% is 4 × $31.50, or $126.

Find the interest on $320.50 :

1. For 4 years at $3\frac{1}{2}$%.
2. For 2 years at 5%.
3. For 3 years at $4\frac{1}{2}$%.

4. For $1\frac{1}{2}$ years at 6%.
5. For $2\frac{1}{4}$ years at 4%.
6. For $1\frac{2}{3}$ years at $5\frac{1}{2}$%.

To find the interest at 6% on $1 for years, months, and days.

The interest at 6% on $1 for 1 year is 6 cents; for 2 yr. is 2 × 6 cents ; for 3 years is 3 × 6 cents ; and so on.

The interest at 6% on $1 for 1 month is $\frac{1}{12}$ of 6 cents ; that is, $\frac{1}{2}$ a cent; for 2 months is 2 × $\frac{1}{2}$ a cent; and so on.

The interest at 6% on $1 for 1 day is $\frac{1}{30}$ of $\frac{1}{2}$ a cent; that is, $\frac{1}{30}$ of 5 mills, or $\frac{1}{6}$ of a mill ; for 2 days is 2 × $\frac{1}{6}$ of a mill ; for 3 days is 3 × $\frac{1}{6}$ of a mill ; and so on. Hence,

We multiply 6 cents by the number of years, $\frac{1}{2}$ a cent by the number of months, and $\frac{1}{6}$ of a mill by the number of days ; and take the sum of these products for the interest on $1 for the given time.

Find the interest at 6% on $1 for 2 yr. 7 mo. 18 dy.

> Int. on $1 at 6% for 2 yr.   = 2 × 6 c.   = $0.12
> Int. on $1 at 6% for 7 mo. = 7 × $\frac{1}{2}$ c.   =   0.035
> Int. on $1 at 6% for 18 dy. = 18 × $\frac{1}{6}$ m. =   0.003
> Int. on $1 at 6% for the given time   = $0.158

Find the interest at 6% on $1 for :

7. 2 years 6 months 6 days.
8. 1 year 7 months 8 days.
9. 1 year 9 months 9 days.

10. 2 years 8 months 2 days.
11. 3 years 5 months 21 days.
12. 2 years 2 months 27 days.

To find the interest at 6% on any principal for years, months, and days:

We find the interest at 6% on $1 for the given time, and multiply this interest by the number of dollars in the principal.

Find the interest at 6% on $213.50 for 2 yr. 7 mo. 18 dy.

The interest on $1 for the given time is $0.158. (See page 226.)

Hence, the interest on $213.50 for the same time and rate is 213.5 × $0.158, and this product is $33.733.

Find the interest on :

    1. $950.50 for 2 yr. 4 mo. 6 dy. at 6%.
    2. $20,000 for 1 yr. 7 mo. 12 dy. at 6%.
    3. $515.25 for 1 yr. 9 mo. 8 dy. at 6%.
    4. $1000 for 2 yr. 1 mo. 19 dy. at 6%.
    5. $216.75 for 2 yr. 2 mo. 21 dy. at 6%.
    6. $927.35 for 1 yr. 8 mo. 28 dy. at 6%.

To find the interest at any other rate than 6%:

We find the interest at 6%, divide this interest by 6, and multiply the quotient by the given rate.

Find the interest on :

    7. $505.90 for 1 yr. 5 mo. 12 dy. at 5%.
    8. $225.40 for 2 yr. 2 mo. 2 dy. at $5\frac{1}{2}$%.
    9. $510.80 for 2 yr. 8 mo. 9 dy. at $4\frac{1}{2}$%.
    10. $2000 for 1 yr. 9 mo. 27 dy. at 4%.
    11. $1200 for 2 yr. 11 mo. 21 dy. at $4\frac{1}{2}$%.
    12. $1500 for 2 yr. 1 mo. 6 dy. at $4\frac{1}{4}$%.
    13. $1600 for 1 yr. 10 mo. 10 dy. at $3\frac{1}{2}$%.
    14. $1300 for 1 yr. 3 mo. 24 dy. at 7%.
    15. $2100 for 3 yr. 4 mo. 12 dy. at $7\frac{1}{2}$%.

BANK DISCOUNT.

A **Note** is a written promise to pay a stated sum of money at a stated time, or on demand.

The stated sum of money is called the **face** of the note; the person who signs the note is the **maker**; the person who writes his name on the back of the note is the **endorser**; the person who has possession of the note is the **holder**.

Banks buy good notes properly endorsed. A bank pays to the holder the amount of the note less the interest on the same for 3 *days more* than the time the note has to run when bought. These 3 extra days are called **days of grace**; and the maker of a note need not pay it to the bank until the *last day of grace*. Interest is reckoned for the exact time the note has to run if the note contains the words *without grace*, or is given in a state which has abolished days of grace by statute.

The money which a bank pays to the holder of a note when the bank buys a note, or *discounts* it as it is called, is the **proceeds** of the note; and the money retained by the bank is the **bank discount**.

Find the proceeds of a 60-day note for $800 with **grace**, discounted at 6%.

The interest on $800 for 63 dy. is $10\frac{1}{2} \times \$0.80 = \$8.40$.
Hence, the proceeds is $\$800 - \$8.40 = \$791.60$.

Find the bank discount and the proceeds of a note without grace:

1. For $250, due in 30 days, discounted at 6%.
2. For $700, due in 4 months, discounted at 6%.
3. For $975, due in 60 days, discounted at $5\frac{1}{2}$%.
4. For $425, due in 2 months, discounted at 5%.
5. For $1100, due in 90 days, discounted at $4\frac{1}{2}$%.
6. For $1200, due in 3 months, discounted at 4%.

REVIEW.

1. In how many days will 24 men do a piece of work that 12 men can do in 10 days?

2. A man took off 5% from a bill for cash payment. If the discount was $36, how much was the whole bill?

3. A man was allowed a discount of $7\frac{1}{2}\%$ on a bill of goods amounting to $370. How much did the goods cost him?

4. Find the cost per pound when 10% is gained by selling butter for 33 cents a pound.

5. A man sold a horse for $210 and lost 30% on the cost. How much did the horse cost him?

6. Find the interest on $324 for 1 yr. 6 mo. 8 dy. at $4\frac{1}{2}\%$.

7. Find the bank discount and the proceeds of a note for $753, due in 4 mo. without grace, discounted at 5%.

8. Find the bank discount and the proceeds of a note for $450, due in 60 days with grace, discounted at $4\frac{1}{2}\%$.

9. Find the amount of $800 at interest for 2 yr. 7 mo. at $5\frac{1}{2}\%$.

10. Find the amount of $420 at interest for 2 yr. at $6\frac{1}{2}\%$.

11. Find the amount of $210 at interest for 1 yr. 2 mo. 3 dy. at 6%.

12. A dealer buys flour at $4 a barrel and sells it for $4.50 a barrel. What per cent does he gain?

13. A commission of $70.84 was charged for selling $2024 worth of wool. What was the rate of commission?

14. The premium for insuring a house, at $\frac{1}{2}\%$, is $15. What is the amount of the insurance?

15. If the discount at 5% on a bill of goods is $25, how much does the purchaser pay for the goods?

BUSINESS FORMS.

(Demand note.)

$300.　　　　　　　Boston, Mass., March 17, 1897.

On demand I promise to pay C. F. Hill, or order, three hundred dollars, with interest at 5%.　Value received.

M. E. Noyes.

(Time note.)

$1250.50.　　　　　　Boston, Mass., March 17, 1897.

Four months after date, I promise to pay J. H. Thayer, or order, twelve hundred fifty and $\frac{50}{100}$ dollars.　Value received.　　　　　　　　　　　　　　　　J. P. Coffin.

(Note without grace.)

$1000.　　　　　　Manchester, N. H., March 16, 1897.

Three months from date, without grace, I promise to pay A. F. Cline, or order, one thousand dollars at the First National Bank.　Value received.　　M. L. Harwood.

(Joint and several note.)

$500.　　　　　　　Nashua, N. H., March 10, 1897.

Ninety days from date, we jointly and severally promise to pay E. L. MacPherson, or order, five hundred dollars at the Second National Bank.　Value received.

E. H. Fairbanks,
C. E. Smith.

(Bank check.)

$310.50.　　　　　　Boston, Mass., March 9, 1897.

Third National Bank,　　　　　　　　No. 24.

Pay to the order of John Hill three hundred ten and $\frac{50}{100}$ dollars.　　　　　　　　　　　　John Templeton.

(Bank draft.)

$500.25.   *Boston, Mass., March 9, 1897.*

Pay to the order of James Drew five hundred and $\frac{25}{100}$ dollars, value received, and charge to account of

To Ninth National Bank,
No. 17.   *New York.* }   *Howard National Bank.*

(Order for money.)

Wakefield, N. H., March 15, 1897.

James Murphy & Co.,

Pay Aaron Young, or order, ten dollars, and charge to my account.   *John Matthews.*

(Order for goods.)

Brookfield, N. H., March 5, 1897.

James Garvin & Co.,

Pay E. F. Hill, or order, ten dollars in goods from your store, and charge to my account.   *H. E. Noyes.*

(Receipt on account.)

$300.   *Concord, N. H., March 5, 1897.*

Received from J. T. Clark three hundred dollars on account.   *John James.*

(Receipt in full.)

$200.   *Brockton, Mass., March 15, 1897.*

Received of John Cotton two hundred dollars in full of all demands to date.   *James Rice.*

(Due-bill for goods.)

Amesbury, Mass., March 19, 1897.

Due J. F. Hill, or order, sixty-five dollars, payable in goods from my store.   *B. F. Moulton.*

MISCELLANEOUS PROBLEMS.

Find the sum of :

1.	2.	3.	4.	5.
236	812	456	845	437
549	298	716	741	553
781	793	310	218	210

6.	7.	8.	9.	10.
884	212	392	872	456
965	566	809	463	789
347	492	318	746	321

11.	12.	13.	14.	15.
429	538	129	822	538
572	763	786	369	465
836	295	543	547	847

16.	17.	18.	19.	20.
196	943	843	565	365
287	868	399	739	.318
735	557	528	208	476

21.	22.	23.	24.	25.
492	387	485	528	367
568	756	968	665	459
279	492	253	789	438

26.	27.	28.	29.	30.
648	567	412	998	862
375	784	866	437	219
594	955	757	569	483

31.	32.	33.	34.	35.
1728	3146	8805	2625	2313
9654	1459	2952	4223	8978
8268	6752	1693	9439	1745
2375	4387	1854	8137	7479

36.	37.	38.	39.	40.
6682	8228	2363	1795	4548
2613	6269	8815	8302	5882
5781	5425	5226	7930	6341
2379	3462	8194	5634	7176

41.	42.	43.	44.	45.
$69.93	$82.58	$61.52	$44.83	$45.28
55.83	42.73	13.24	79.68	30.84
68.10	52.65	98.77	48.17	37.85
60.63	53.56	28.94	84.78	98.62
57.75	48.32	79.97	63.43	54.87

46.	47.	48.	49.	50.
$84.37	$59.24	$59.90	$16.78	$48.62
85.98	73.77	40.29	15.24	52.86
62.24	26.13	68.58	72.45	17.65
84.26	68.65	58.20	63.04	89.59
95.48	23.58	83.38	86.69	74.29

51.	52.	53.	54.	55.
$75.95	$57.96	$26.58	$57.82	$42.54
76.22	46.17	92.19	25.51	71.43
78.47	80.36	19.68	36.84	44.67
13.78	13.95	63.15	46.52	75.46
47.74	58.64	59.18	72.93	28.29

56.	57.	58.	59.	60.
5498	6497	5847	6892	5816
8729	9216	2805	4282	1476
4298	1728	4796	5734	8817
6542	1331	5277	7992	4444
2817	8725	8429	2498	6575

61.	62.	63.	64.
$129.78	$426.58	$1728.48	$ 719.63
64.55	742.74	231.25	2422.58
49.42	875.63	14.79	3785.71
276.54	587.21	586.47	1879.47
52.99	215.40	2268.55	7956.80
622.48	653.39	418.73	1919.66

65.	66.	67.	68.
233,629	617,024	849,375	682,929
245,625	481,325	417,954	479,236
177,450	539,148	349,354	518,648
121,987	734,986	288,942	565,787
593,640	240,608	219,723	653,456
191,478	369,391	593,640	324,475

69.	70.	71.	72.
928,754	472,618	568,896	327,412
547,488	197,564	423,732	945,475
695,849	219,858	617,469	889,729
754,955	343,373	755,237	678,857
876,576	909,909	822,545	796,668
638,737	188,786	576.678	517,543

73.	74.	75.	76.
$2718.28	$323.58	$1543.17	$ 848.67
402.56	821.90	2023.31	1929.75
189.73	658.13	737.75	2887.44
6897.65	827.94	1892.64	25.25
172.85	939.33	3141.68	428.72
675.34	711.56	2772.95	533.69

77.	78.	79.	80.
141.82	4.9287	0.1728	168.729
62.975	5.2548	3.1416	5.2368
0.428	17.3683	0.7854	74.587
77.65	2.9875	2.9667	4.9547
82.489	0.0028	8.5775	0.6285
163.8	0.1497	9.4386	615.24

81.	82.	83.	84.
314.8276	42.7869	0.6184	491.205
9.8723	2.236	52.92	68.325
41.9487	17.3258	0.7928	4.9872
72.9434	0.6213	0.2874	0.5777
6.1348	15.4325	62.5	0.0752
42.7869	3.7859	4.8655	68.5496

85.	86.	87.	88.
88.9728	0.3246	67.428	23.8995
38.3435	25.827	119.5687	85.6374
4.3433	234.19	48.9276	53.7189
0.5799	48.684	0.5236	90.094
679.4287	0.7866	17.2845	96.3438
85.6378	57.6754	4.6338	7.4817

Find the remainder and prove:

89. 5388 − 4792.

90. 5000 − 2896.

91. 6000 − 4895.

92. 82,336 − 36,798.

93. 85,004 − 76,584.

94. 90,005 − 27,129.

95. 91,617 − 76,824.

96. 47,819 − 28,927.

97. 56,427 − 49,849.

98. 82,315 − 43,768.

99. 62,134 − 28,456.

100. 84,718 − 48,626.

101. 98,732 − 87,654.

102. 61,830 − 38,196.

103. 44,712 − 28,680.

104. 72,943 − 61,348.

105. 42,786 − 42,595.

106. 23,606 − 6180.

107. 47,892 − 19,736.

108. 32,808 − 16,093.

109. 78,539 − 30,479.

110. 82,618 − 47,891.

111. 56,487 − 48,714.

112. 89,173 − 35,428.

113. 19,001 − 3872.

114. 77,000 − 48,782.

115. 6855 − 4728.5.

116. $87.92 − $48.64.

117. $119.28 − $78.59.

118. $162.49 − $88.75.

119. $647.51 − $549.64.

120. $270.04 − $138.37.

121. $247.93 − $128.58.

122. $549 − $418.72.

123. 0.5287 − 0.0479.

124. 9.873 − 8.1596.

125. 5.8207 − 2.2643.

126. 10 − 4.7286.

127. 19.26 − 15.7542.

128. 6532.18 − 1916.47.

129. $262.18 − $1.25.

130. 1 − 0.42865.

131. 10.074 − 6.2854.

132. 21,205 − 16,827.48.

133. 481.796 − 219.487.

134. 3.785 − 0.62137.

135. 3.1416 − 2.7182.

136. $42.85 − $28.67.

137. 58.263 − 42.7854.

138. 204.01 − 85.006.

139. 1000 − 91.682.

140. 24.503 − 8.96.

141. 0.006 − 0.0042.

142. 1.896 − 0.9548.

143. 39.3704 − 39.1392.

144. 13.0003 − 8.5096.

145. 478.19 − 326.189.

146. $526.92 − $475.15.

147. 114 − 94.5868.

148. 62.809 − 56.4293.

149. 15.7862 − 15.7287.

150. 70.937 − 45.8643.

151. 125.4 − 75.5986.

152. 46.7 − 46.528.

Find the product of:

153. 8 × 1728.
154. 9 × 2654.
155. 24 × 8268.
156. 47 × 5682.
157. 56 × 4297.
158. 31 × 68.48.
159. 121 × 495.
160. 267 × 548.
161. 358 × 962.
162. 789 × 826.
163. 1416 × 2865.
164. 2762 × 3548.
165. 3589 × 4608.
166. 2806 × 5479.
167. 3006 × 6008.
168. 14.9 × 742.
169. 25.7 × 689.
170. 36.5 × 860.
171. 60.8 × 1492.
172. 0.187 × 2687.
173. 0.54 × 0.618.
174. 0.49 × 0.785.
175. 1.62 × 0.847.
176. 26.4 × 26.4.
177. 3.08 × 74.64.
178. 204.01 × 168.55.
179. 36 × 24.869.
180. 98 × 47.862.
181. 1001 × 4699.
182. 354 × 5280.
183. 476 × 452.
184. 250 × 867.

185. 2000 × 1486.
186. 3600 × 746.
187. 468 × 8035.
188. 807 × 3063.
189. 418 × 9876.
190. 0.592 × 0.618.
191. 0.455 × 62.
192. 0.078 × 4176.
193. 0.029 × 592.6.
194. 0.0061 × 6159.
195. 42.28 × 148.
196. 9.786 × 536.5.
197. 498.2 × 3.682.
198. 536.78 × 1492.
199. 48.75 × 58.6.
200. 0.025 × 82,768.
201. 0.375 × 60,780.
202. 0.83 × 56,812.
203. 0.789 × 0.685.
204. 0.809 × 15.65.
205. 0.065 × 288.
206. 0.049 × 7.684.
207. 32.9 × $49.87.
208. 37.6 × $67.39.
209. 9.87 × $28.75.
210. 14.62 × $186.75.
211. 54.96 × 54.96.
212. 87.35 × 636.89.
213. 0.0497 × 0.543.
214. 5875 × 6297.
215. 6.1587 × 1728.
216. 5.1974 × 0.0934.

Find the quotient of :

217. 1728 ÷ 12.
218. 8789 ÷ 43.
219. 9988 ÷ 97.
220. 6548 ÷ 68.
221. 89,713 ÷ 79.
222. 60,512 ÷ 82.
223. 43,657 ÷ 38.
224. 37,465 ÷ 35.
225. 84,793 ÷ 62.
226. 43,269 ÷ 64.
227. 67,540 ÷ 101.
228. 59,736 ÷ 205.
229. 94,362 ÷ 409.
230. 46,327 ÷ 864.
231. 834,561 ÷ 189.
232. 341,586 ÷ 498.
233. 861,345 ÷ 843.
234. 370,406 ÷ 395.
235. 978,217 ÷ 357.
236. 604,730 ÷ 248.
237. 456.25 ÷ 1.25.
238. 492 ÷ 7.18.
239. 546 ÷ 8.46.
240. 687 ÷ 953.
241. 327.5 ÷ 0.025.
242. 0.0125 ÷ 2.5.
243. 300 ÷ 3.125.
244. 15 ÷ 0.205.
245. 665.25 ÷ 500.
246. 1050 ÷ 4.26.
247. 8236 ÷ 5.504.
248. 46.87 ÷ 518.2.

249. 68,719 ÷ 3003.
250. 46,435 ÷ 35,000.
251. 0.06 ÷ 0.518.
252. 0.09 ÷ 4.62.
253. 12,876 ÷ 33,000.
254. 640 ÷ 16,000.
255. 12.5 ÷ 0.375.
256. 0.625 ÷ 5.19.
257. 48.37 ÷ 4.682.
258. 33.512 ÷ 0.025.
259. 962.5 ÷ 0.625.
260. 587.62 ÷ 0.375.
261. 0.55 ÷ 0.0011.
262. 5.92 ÷ 23.03.
263. 309.45 ÷ 15,000.
264. 68.47 ÷ 712.
265. 9.006 ÷ 0.568.
266. 10 ÷ 0.625.
267. 1728 ÷ 0.75.
268. 1486 ÷ 0.125.
269. 405.15 ÷ 0.222.
270. 327.5 ÷ 0.025.
271. 75 ÷ 0.025.
272. 1050 ÷ 43.75.
273. 67 ÷ 5000.
274. 58.6728 ÷ 0.58.
275. 46.872 ÷ 4.12.
276. 56.935 ÷ 5.685.
277. 15 ÷ 0.4872.
278. 1667 ÷ 4.5987.
279. 136 ÷ 1.7985.
280. 149 ÷ 22.7638.

ADDITION.

281. Add four thousand six hundred forty-seven, two thousand nine hundred seventy-nine, eighteen hundred ninety-nine, eight thousand four hundred twenty-three.

282. Add sixty-eight and seventy-two hundredths, one hundred forty-seven and thirty-six hundredths, two and forty-one hundredths, ninety and four hundredths.

283. A man drew $86.25 from a bank and had $642.35 left. How much did he have in the bank at first?

284. A lumber dealer sawed from one lot 42,694 feet of boards, 68,712 feet from a second, 57,982 feet from a third, 82,576 feet from a fourth. How many feet of boards did he saw from the four lots?

285. A farm contains 96.26 acres of woodland, 72.78 acres of pasture, 48.46 acres of tillage land, and 18.58 acres of meadow. How many acres does the farm contain?

286. A man paid for a horse $175, for a carriage $115, for a harness $42, and for a whip $1.50. How much did he pay for them all?

287. The area of England is 50,535 square miles; of Scotland, 29,167 square miles; and of Wales, 8125 square miles. Find the area of Great Britain.

288. A man was born in 1868. When will he be 64 years old?

289. A man paid $6250 for a house and lot, and spent $287.50 for repairs, $115.25 for painting, and $152.90 for grading. What was the total cost of the house and lot?

290. In 1897, Iowa raised 220,089,149 bushels of corn; Nebraska, 241,268,490 bushels; Kansas, 162,442,728 bushels. How many bushels did the three states raise?

291. In 1890 the population of Maine was 661,086; of New Hampshire, 376,530; of Vermont, 332,422; of Massachusetts, 2,238,943; of Rhode Island, 345,506; of Connecticut, 746,258. Find the total population of New England.

292. In 1890 the number of school children in Maine was 201,851; in New Hampshire, 106,611; in Vermont, 101,457; in Massachusetts, 650,870; in Rhode Island, 105,534; in Connecticut, 221,245. Find the total number of school children in New England.

293. The area of Maine is 33,039 square miles; of New Hampshire, 9377 square miles; of Vermont, 9563 square miles; of Massachusetts, 8546 square miles; of Rhode Island, 1247 square miles; of Connecticut, 5612 square miles. Find the total area of New England.

294. The scores of the five men of a bowling team are respectively 542, 497, 486, 538, 519. What is the total score of the team?

295. A builder contracted to build four houses for respectively $8247.60, $6575.50, $6142.35, $5642.65. What is the total amount of his contracts?

296. The sales of a store on Monday were $424.67, Tuesday $362.48, Wednesday $477.88, Thursday $441.84, Friday $497.62, Saturday $682.27. What was the total amount of sales for the week?

297. A storekeeper pays $750 a year for rent, $832 for one clerk, $728 for a second, $326.50 for fuel and lights, and $127.83 for other expenses. What is the expense of running his store for a year?

298. Four men form a partnership. They furnish respectively $7527, $2428, $3216, and $5489. What is the amount of their capital?

299. Four loads of hay weighed respectively 2264 pounds, 1989 pounds, 1867 pounds, and 2182 pounds. What was the total weight of the hay?

300. The main line of a railroad is 382.16 miles long. There are five branches whose lengths are respectively 12.82 miles, 64.27 miles, 4.19 miles, 28.96 miles, and 7.78 miles. What is the entire length of the railroad?

301. In 1898 the internal revenue receipts of the United States were as follows: spirits $99,283,534, tobacco $52,493,208, fermented liquors $68,644,558, stamps $43,837,819, legacies $1,235,435, special taxes $4,921,593, miscellaneous $3,068,425. What were the total internal revenue receipts?

302. The total expenses of the United States for the years 1789–1899 were: civil 3162 million dollars, war 5401 million, navy 1513 million, Indians 358 million, pensions 2518 million, interest 2942 million. Find the number of millions spent.

303. Lake Superior has 32,290 square miles of surface, Lake Michigan 23,903, Lake Huron 23,684, Lake Erie 9439, Lake Ontario 7654. Find the total area of these lakes in square miles.

304. A man had in a bank the first day of a month $1628.47. He deposited in the bank at different times during the month $462.85, $722.49, $86.72, $146.23. What was his balance at the end of the month?

305. The smaller of two numbers is 14.167, and the difference between the numbers is 47.872. Find the larger.

306. In 1895 the value of the exports of the United States was $807,538,165, and the value of the exports in 1899 was $419,485,137 greater. Find the value in 1899.

307. A man sold three bonds.   For the first he received $1225, for the second $990, and for the third $1375. What did he receive for the three bonds ?

308. An orchard contains 347 apple trees, 286 pear trees, 196 peach trees, and 47 cherry trees.   How many trees are there in the orchard ?

309. A butcher collects $147.56 from one customer, $28.75 from a second, $69.43 from a third, and $44.78 from a fourth.   What is the amount of his collections ?

310. A farmer threshed 46.8 bushels of wheat, 34.6 bushels of oats, and 64.8 bushels of beans.   How many bushels did he thresh in all?

311. In 1890 the population of New York City was 1,515,301, and that of Brooklyn was 806,343.   What was the population of the two cities together ?

312. A bank had $54,782 in gold, $82,784 in bills, and $16,412.82 in other cash items.   Find the whole account.

313. A cattle dealer has sold 42 oxen, 87 sheep, and 37 horses; he has left 146 oxen, 529 sheep, and 187 horses. How many animals did he have at first ?

314. The difference between two cargoes of coal is 3754.09 tons and the smaller cargo has 2789.11 tons.   Find the number of tons in the larger cargo.

315. The army of the Duke of Wellington at Waterloo consisted of infantry 20,661, cavalry 8735, artillery 6877, and allies 33,413.   What was the strength of his army ?

316. The army of Napoleon at Waterloo consisted of 48,950 infantry, 15,765 cavalry, and 7732 artillery.   What was the strength of his army ?

317. Find the number that is larger than 14,782.82 by the sum of 768 and 1548.791.

SUBTRACTION.

318. Subtract five thousand seven hundred sixty-nine from eight thousand two hundred thirty-six.

319. Subtract forty-two and thirty-four hundredths from eighty-seven and twelve hundredths.

320. A farmer owned 642.4 acres of land and sold a pasture of 84.7 acres. How many acres did he still own?

321. A merchant owed a note of $2672.50 and made a payment of $1128.75. How much did he still owe?

322. The population of the United States in 1890 was 62,622,250, and in 1880 was 50,155,783. What was the gain in population from 1880 to 1890?

323. A farmer raised 5712 bushels of wheat and sold 5284 bushels. How many bushels of wheat did he have left?

324. An agent bought a lot for $12,672.50 and sold it for $15,284. How much did he gain by the transaction?

325. A man had an income of $1852.65 and his expenses were $1476.35. How much did he save during the year?

326. A father bequeathed to his daughter $10,625 and to his son $6778.25. How much more did he bequeath to his daughter than to his son?

327. A merchant owned stock to the value of $12,482.84. The stock was damaged by a fire to the amount of $6847.62. What was then the value of the stock?

328. A ship worth $22,862 was insured for $16,425.50. If the ship was totally destroyed by a wreck, what was the loss of the owner?

329. Two men are together worth $38,472.45. If the first is worth $22,879.85, what is the amount the second is worth?

330. The sum of two numbers is **15,** and the smaller is 6.847. Find the larger.

331. The sales of a store amounted to $12,847.62 for a year. The cost of the goods was $8972.45 and the expenses of running the store were $1246.82. Find the profit for the year.

332. What number must be added to 19.212 to make 30 ?

333. A man sold a city lot for $4875, and bought a farm for $2289.75. How much did he have left ?

334. How many years from the discovery of America in 1492 to the Declaration of Independence in 1776 ?

335. The area of France is 204,971 square miles and that of California is 188,981 square miles. How many square miles is France larger than California ?

336. If 168.41 is the minuend and 94.63 the subtrahend, what is the remainder ?

337. The distance from New York to Chicago is 982.24 miles and the distance from New York to Buffalo is 441.75 miles. Find the distance from Buffalo to Chicago.

338. At sunset the mercury of a thermometer stood at 52.8° and at sunrise at 34.6°. How many degrees did it fall during the night ?

339. Before a storm the mercury in a barometer fell from 30.308 inches to 29.746 inches. How many inches did it fall ?

340. How much must be added to nine thousandths to make the whole number nine ?

341. One pound of dry oak wood when burnt yields 0.022 pound of ashes. What part of the pound disappears in the air ?

342. In 1890 Minneapolis had 164,738 inhabitants and St. Paul 133,156. How much larger was Minneapolis?

343. In 1890 New York had 1,515,301 inhabitants and Chicago 1,099,850. How much larger was New York?

344. The area of Lake Superior is 32,290 square miles and that of Lake Erie is 9493 square miles. How many square miles greater is the area of Lake Superior than that of Lake Erie?

345. In an army of 12,682 men, 77 were killed, 489 were wounded, and 329 were missing. How many men could the army then muster?

346. A lumber dealer had 786,000 feet of boards and sold 298,422 feet. How many feet of boards did he still have?

347. If two candidates for governor received together 142,608 votes and if the successful candidate received 76,489 votes, how many votes were received by the defeated candidate?

348. A man paid $872.65 for a pacing horse and sold him for $1062.40. How much did he gain?

349. A cistern that will hold 122.75 barrels contains 78.9 barrels of water. How many barrels of water will it take to fill the cistern?

350. The receipts of a shoe factory for a year were $1,648,762.25 and the expenses were $1,594,287.60. Find the profit.

351. Two men start from the same place and travel in the same direction. One travels 595.25 miles and the other 789.01 miles. How far apart were they then?

352. How many years from the discovery of America in 1492 to the year 1900?

353. In 1889 the number of pupils that took Latin in the preparatory schools of the United States was 100,144. In 1897 the number was 274,293. Find the increase.

354. In 1889 the number of pupils that studied algebra was 127,397. In 1897 the number had increased to 306,755. Find the increase.

355. The receipts of the United States government in 1893 were $385,818,629 and the expenditures were $383,-477,954. Find the excess of receipts over expenditures.

356. A merchant bought goods for $9472.64 and sold them for $11,228.42. How much did he gain?

357. Two men are 2712 miles apart. They travel towards each other, one 569 miles and the other 786 miles. How many miles are they then apart?

358. A contractor built a bridge for $21,407.65 and the work cost him $19,872.50. What was his profit?

359. A commission merchant bought 14,475 bushels of wheat and sold 4480 bushels and then 5275 bushels. How many bushels of wheat did he then have?

360. A man sold a city lot for $18,712.75 and bought a farm for $12,488. How much did he have left?

361. A depositor has $12,817.63 in a bank and draws out $8294.86. How much has he left on deposit?

362. The value of a ship and cargo is $1,320,592.60 and the value of the cargo is $482,728.60. What is the value of the ship?

363. The polar diameter of the earth is 41,707,620 feet and the equatorial diameter is 41,847,426 feet. What is the difference?

MULTIPLICATION.

364. Multiply eighteen hundred sixty-three by forty-two.

365. Multiply seventy-four and thirty-two hundredths by five and seventy-three hundredths.

366. Multiply eight hundred fifty-seven by six hundred seventeen thousandths.

367. A drover bought 114 sheep at $3.75 apiece. How much did he pay for the sheep?

368. If a clerk receives $65 a month, how much does he receive for a year?

369. Find the cost of 179 acres of land at $17 an acre.

370. Find the cost of 628 baskets of peaches at $1.75 a basket.

371. Find the cost of 144 barrels of flour at $5.25 a barrel.

372. If a carpenter earns $56.25 a month, how much does he earn in a year?

373. If sound travels 1152 feet a second, how far will it travel in 28 seconds?

374. Find the cost of 8.25 yards of cloth at $0.75 a yard.

375. A railroad 63 miles long cost $12,782.35 a mile. What was its entire cost?

376. A clock strikes 156 times a day. How many times does it strike in a year of 365 days?

377. If an acre of corn is worth $33.35, what is the value of 583 acres of corn?

378. The cargo of a ship consists of 784 boxes of oranges, each box containing 249 oranges. How many oranges are there in the cargo?

379. If a speaker utters 128 words a minute, how many words does he deliver in a speech of 17 minutes?

380. A party of sixteen hired a schooner for a cruise and paid $16.35 apiece. How much did all together pay?

381. In an orchard there are 46 rows of trees and 34 trees in a row. How many trees in the orchard?

382. The earth moves in its orbit at the rate of 19 miles a second. How many miles does it move in one minute? How many miles in one hour?

383. A dealer bought 24 firkins of butter, each firkin containing 48 pounds, at $0.18 a pound. How much did he pay for the butter?

384. If a mill turns out 268.75 tons of steel rails in a day, how many tons of rails will it produce in 93 days?

385. What is the cost of 96.85 reams of paper at $3.185 a ream?

386. How many yards are there in 38 pieces of carpeting, if each piece contains 46.25 yards?

387. There are 2150.42 cubic inches in a bushel. How many cubic inches in 57 bushels?

388. In building a hall 1,576,000 bricks were used. Find the cost of the bricks at $6.75 a thousand.

389. The distance round a circle is 3.1416 times the distance across it. If the distance across a circle is 17.6 inches, find the distance round it.

390. The distance across a circle is 0.31831 of the distance around it. Find the distance across a bowling ball if the distance round it is 27 inches.

391. If gas costs $2.15 a thousand cubic feet, find the amount of the bill for 8640 cubic feet of gas.

392. A grocer bought 598 pounds of sugar at $0.05125 a pound, and sold it at $0.055 a pound.   Find his profit.

393. A workman receives $2.75 a day and spends on the average $1.90 a day.   How much does he save in a year of 365 days, if he rests 52 Sundays and 4 holidays?

394. If milk is 1.03 times as heavy as water and a cubic foot of water weighs 62.5 pounds, what is the weight of a cubic foot of milk?

395. Find the distance from the earth to the sun, if light travels 186,000 miles a second, and it requires 493 seconds for light from the sun to reach the earth?

396. A merchant has 79 barrels of sugar, each weighing 287 pounds.   How many pounds of sugar has he?

397. A school is composed of 68 boys whose average weight is 129 pounds.   Find the total weight of the boys.

398. A farmer sold 18 cows for $27.75 apiece.   How much did he receive?

399. A butcher sold 547 pounds of bacon at 13 cents a pound.   How much did he receive?

400. In 1897 New York produced 15,335,142 bushels of corn.   Find its value at 41 cents a bushel.

401. In 1897 Indiana produced 32,675,201 bushels of wheat.   Find its value at $0.915 a bushel.

402. A tree dealer sold 3246 young fruit trees at 14 cents a tree.   How much did he receive for them?

403. How many persons die in a year in a city of 192,000 inhabitants if the annual death rate is 19 per thousand?

404. If an acre of corn is worth $26.85, what is the value of 2106 acres of corn?

405. A cubic foot of hard coal weighs 84 pounds. How many pounds will 29 cubic feet weigh?

406. If an oil well produces 337.75 barrels of oil in a day, how many barrels of oil will it produce in 27 days?

407. A certain number divided by 342 gives 1471 for the quotient and 112 for the remainder. Find the number.

408. The circumference of a circle is 3.1416 times its diameter. What is the circumference of a circle whose diameter is 27 inches?

409. A farmer sold 0.375 of his farm of 248 acres. How many acres had he left?

410. A boy can point 13,427 pins an hour. How many pins can he point in 6 days of 8 hours each?

411. The area of Massachusetts is 8546 square miles and in 1890 the average number of inhabitants per square mile was 262. What was the population?

412. A grocer sold 6790 pounds of sugar at $4.85 a hundred pounds. How much did he receive?

413. A lumber dealer sold 17,925 feet of boards at $16.50 a thousand feet. How much did he receive?

414. A drover bought 375 lambs at $3.75 a head. How much did he pay for his flock?

415. Of what number is 1728 both the divisor and the quotient?

416. In 1899 there were 39,114,453 sheep in the United States, and their average value was $2.75. What was the total value of the sheep?

417. The value of the pound Sterling is $4.8665 in United States money. What is the value of 1800 pounds?

DIVISION.

418. Divide eleven thousand one hundred twelve by four hundred sixty-three.

419. Divide two hundred fifteen and thirty-four hundredths by thirty-seven.

420. Divide one hundred seventy-six and four hundred sixty-two thousandths by twenty-eight and six hundredths.

421. How many tons of coal at $6 a ton can be bought for $1344 ?

422. An express train ran 546 miles in 13 hours. What was the rate per hour ?

423. How many shares of mining stock at $39 a share can be bought for $3042 ?

424. If a man travels 27 miles a day, how many days will it take him to travel 621 miles ?

425. The product of two numbers is 40.32. If one of the numbers is 14.4, what is the other ?

426. If a town containing 57,920 acres is divided into farms of 160 acres each, what will be the number of the farms ?

427. How many yards of cloth at $0.125 a yard can be bought for $45.50 ?

428. A hat factory can make 350 hats a day. How many days will it take to fill an order of 28,000 hats ?

429. How many times can 239 be subtracted from 4302 ?

430. What number multiplied by 79 will give the same product as 279 multiplied by 553 ?

431. How many street cars, each containing 72 persons, will be required to carry 3240 persons ?

432. A carpenter received $63 for 36 days' work. How much did he receive a day?

433. If a field of 238 acres produced 8568 bushels of wheat, what was the average yield per acre?

434. At $12.50 a ton, how many tons of hay can be bought for $225?

435. The dividend is 3856, the quotient 142, and the remainder 22. Find the divisor.

436. A ship sailed from Liverpool to Cape Town, 7000 miles, in 25 days. What was her average daily rate?

437. If sound travels 1060 feet a second, and a cannon is fired at a distance of 16,960 feet, how many seconds after the flash will the report be heard?

438. A man dug a ditch for $225 in 125 days. How much did he receive a day?

439. A cider merchant made 12,000 gallons of cider, which he put into casks containing 42 gallons each. How many full casks did he have, and how many gallons over?

440. If it costs $0.25 a word to cable a message to England, how many words can be cabled for $8.25?

441. How many acres of land can be bought for $931, at $49 an acre?

442. If 19 oxen cost $1330, and an ox costs as much as 14 sheep, what is the cost of a sheep?

443. The attendance at a school for one week was as follows: Monday 116, Tuesday 109, Wednesday 85, Thursday 96, Friday 114. What was the daily average attendance for the week?

444. By what number must 5376 be multiplied to make 6,521,088?

445. Of what number is 289 both divisor and quotient?

446. The heights of a barometer on four successive mornings were respectively 29.743 inches, 29.786 inches, 29.697 inches, and 29.74 inches. What was the average height?

447. A man bought a farm for $10,000. If the buildings are worth $2400 and the farm contains 190 acres, how much per acre did he pay for the land?

448. How many pairs of trousers, each pair requiring 2.375 yards, can be made from a piece of cloth that contains 33.25 yards?

449. The imperial gallon of Great Britain contains 277.274 cubic inches. If a gallon contains 8 pints, how many cubic inches are there in an imperial pint?

450. A drover sold 27 hogs for $236.25. How much did he receive for each?

451. How many freight cars, costing $385 each, can be bought for $18,480?

452. At $22.25 an acre, how many acres of land can be bought for $1379.50?

453. A bushel of corn weighs 56 pounds. How many bushels are there in a carload of corn that weighs 29,568 pounds?

454. In how many hours will a pipe that discharges 107.5 gallons an hour empty a cistern that contains 2580 gallons of water?

455. If the distance of the moon is 240,000 miles and that of the sun is 92,500,000 from the earth, how many times as far is it to the sun as to the moon?

456. The salary of the president of the United States is $50,000 a year. How much does he receive each day?

457. A freight train loaded with flour carries 2812 barrels and each car contains 148 barrels. How many cars are there in the train?

458. If 3.375 yards of cloth are required to make a coat, how many coats can be made from 94.5 yards of cloth?

459. If a farmer raised 4468 bushels of corn from 84 acres, how many bushels did each acre produce?

460. The cost of a railroad was $38,421,762.50. What was the average cost per mile if the road was 867 miles long?

461. How many times can 886 be subtracted from 48,762 and what will be the remainder?

462. The area of British India is 1,004,616 square miles and the population 150,767,851. How many inhabitants are there to a square mile?

463. There are 640 acres in a square mile. How many square miles are there in the District of Columbia, which contains 38,400 acres?

464. The greatest height ever reached by a balloon is 37,000 feet. A mile is 5280 feet. How many miles has a balloon ascended?

465. A farmer put 0.375 of his oats into bins and sold the rest. How many bushels did he sell, if he put 480 bushels into bins?

466. In Iowa, in 1897, 220,089,149 bushels of corn were raised. The area planted was 7,589,281 acres. What was the average yield per acre in bushels?

467. In 1897, 29,907,392 bushels of oats were raised in Ohio. The number of acres sown was 934,606, and the total value of the oats was $5,981,478.40. What was the average yield per acre and the average value per bushel?

COMPOUND QUANTITIES.

468. How many bushels are there in 4920 lb. of potatoes ?

469. How many bushels are there in 3072 lb. of oats ?

470. How many bushels are there in 6608 lb. of corn ?

471. How many grains are there in 4 ingots of silver, each weighing 17 oz. 11 dwt. ?

472. A field is divided into 24 gardens, each containing 25 sq. rd. 16 sq. yd.   What is the area of the field ?

473. If a man takes 720 steps of 2 ft. 10 in. each in 5 min., how far will he walk in an hour ?

474. How many bottles, each holding 1 pt. 3 gi., can be filled from a cask of wine that contains 114 gal. ?

475. Washington was born Feb. 22, 1732, and died Dec. 14, 1799.   How old was he when he died ?

476. Napoleon was born Aug. 15, 1769, and died at the age of 51 yr. 8 mo. 20 dy.   What was the date of his death ?

477. If telegraph posts are placed 88 yd. apart, and a passenger notices that one is passed every 3 seconds, how many miles an hour is the train travelling ?

478. From a field containing 22 A. 105 sq. rd. 18 sq. yd. a farmer sold 8 A. 145 sq. rd. 20 sq. yd.   How much did he have left ?

479. If a milkman sells 121 qt. of milk a day, how many gallons will he sell in a month of 30 dy. ?

480. What will be the cost of 54,630 lb. of wheat at 75 cents a bushel ?

481. If a man saves $77.40 from a yearly salary of $1260, how much does he spend a day ?

482. The length of the sidereal year is 365.2564 days. What is the length in days, hours, minutes, and seconds?

483. The length of the lunar month is 29.53 days. What is the length in days, hours, minutes, and seconds?

484. What fraction of 21 cu. yd. 11 cu. ft. 1215 cu. in. is 3 cu. yd. 1 cu. ft. 1161 cu. in.?

485. The moon revolves round the earth in 27 dy. 7 hr. 43 min. Express this interval of time in days.

486. How many quart boxes will be needed to hold 8 bu. 3 pk. 5 qt. of blueberries?

487. In three piles of wood there are respectively 12 cd. 48 cu. ft., 16 cd. 75 cu. ft., and 18 cd. 37 cu. ft. What is the value of the three piles at $4.25 a cord?

488. A firkin of butter weighs 42 lb. 5 oz., and the firkin weighs 6 lb. 10 oz. What is the weight of the butter?

489. How many times does a bicycle wheel 7 ft. 4 in. in circumference turn in going 3 mi.?

490. How many pounds avoirdupois does a cubic yard of water weigh, if a cubic foot weighs 1000 oz.?

491. Reduce 0.455 bu. to quarts.

492. An imperial gallon of water weighs 10 lb. and a cubic foot of water weighs 1000 oz. How many imperial gallons are there in a cistern that contains 128 cu. ft.?

493. A wine merchant had 236 gal. 2 qt. 1 pt. of wine and sold 154 gal. 3 qt. How much wine did he have left?

494. If a man bought 5 reams of paper, how many sheets did he buy?

495. If a merchant bought 500 gross of pens, how many pens did he buy?

BILLS.

Make out receipted bills for the following accounts, supplying dates :

496. John H. Fellows bought of A. S. Langley 17 lb. of ham at 14 cents, 32 lb. of beefsteak at 24 cents, 28 lb. of mutton at $12\frac{1}{2}$ cents, 16 lb. of veal at 14 cents, $22\frac{1}{2}$ lb. of chickens at 17 cents, and 36 dozen of eggs at 28 cents.

497. John Brown bought of Hilliard & Kimball 32 lb. of butter at 32 cents, 12 lb. of coffee at $37\frac{1}{2}$ cents, 14 lb. of lard at 9 cents, 50 lb. of sugar at 5 cents, 2 lb. of tea at 70 cents, and 7 gal. of molasses at 35 cents.

498. W. P. Chadwick bought of Harry Anderson 12 tons of furnace coal at $5.75, 5 tons of stove coal at $6.25, 4 cd. of hard wood at $4.50, $2\frac{1}{2}$ cd. of pine wood at $3.00, and 2 loads of kindlings at 75 cents.

499. Albert Dow bought of Frank H. Wadleigh 35 bbl. of apples at $2.75, $4\frac{1}{2}$ tons of hay at $18, 25 bu. of potatoes at 75 cents, 3 bu. of onions at 90 cents, and 50 bu. of corn at 80 cents.

500. R. P. Thompson sold to Daniel Gilman 4 bbl. of flour at $5.25, 1 bbl. of sugar, 248 lb., at $5\frac{1}{8}$ cents a pound, 112 lb. of butter at 25 cents, 48 dozen eggs at 22 cents, 28 lb. of coffee at 35 cents, 16 lb. of cheese at 12 cents, and 20 bu. of potatoes at 90 cents, and took in exchange 8 cd. of birch wood at $4.25 and 6 cd. of pine wood at $2.75.

501. George L. Nason sold to George Richards 12 bags of cracked corn at 85 cents, 8 bags of oats at 80 cents, 200 lb. of wire nails at $3\frac{1}{2}$ cents, 50 lb. of sweet potatoes at 3 cents, 2 bu. of Baldwin apples at 60 cents, 8 gal. of maple syrup at $1.00, and 12 bu. of pears at 75 cents.

MEASUREMENTS.

502. The diameter of a bicycle wheel is 28 inches. What is its circumference ?

503. How many rods of fence are required to enclose a rectangular field 40 rods long and 25 rods wide ?

504. Find the area of a rectangular field 260 yards long and 225 yards wide.

505. How many bricks, 9 in. long and 4 in. wide, will be needed to pave a court 24 ft. square ?

506. Find the volume of a rectangular cistern 12 ft. long, 9 ft. wide, and 7 ft. high.

507. Find the area of a circle 14 inches in diameter.

508. How far has a carriage gone when its fore wheels, $3\frac{1}{2}$ ft. in diameter, have each made 862 revolutions ?

509. A rectangular field, 20 rd. wide, contains 3 A. 20 sq. rd. What is the length of the field ?

510. A rectangular cistern, 9 ft. wide and 6 ft. high, contains 702 cu. ft. What is the length of the cistern ?

511. If a carriage wheel makes 440 revolutions in going a mile, what is the diameter of the wheel ?

512. How many cubic feet of stone are required to build a wall 88 ft. long, 25 ft. wide, and 6.5 ft. high ?

513. Around a circular pond, 35 ft. in diameter, is a walk 7 ft. wide. What will it cost to gravel the walk at 25 cents a square yard ?

514. How many cords of wood are there in a pile 80 ft. long, 6 ft. high, and 4 ft. wide ?

515. What is the cost of a pile of wood 36 ft. long, 5 ft. high, and 4 ft. wide at $3.75 a cord ?

516. How many acres are there in a field 360 yards long and 121 yards wide?

517. How many yards of carpeting 27 in. wide will be required for a room 15 ft. by 13½ ft., if the strips run lengthwise? How many if the strips run across the room?

518. Find the cost of carpeting 27 in. wide, at $2.50 a yard, for a room 17 ft. 4 in. by 14 ft. 3 in., if the strips run lengthwise and if there is a waste of ¼ yd. a strip in matching the pattern.

519. Find the cost of carpeting 30 in. wide, at $1.75 a yard, for a room 17 ft. by 14 ft., if the strips run across the room.

520. How many double rolls of paper will be required for a room 15 ft. by 12 ft., if the room has one door 4 ft. wide, and three windows, each 3½ ft. wide?

521. At $1.25 a single roll, put on, what is the cost of papering a room 18 ft. 6 in. long and 16 ft. wide, if the room has two doors, each 3 ft. 9 in. wide, and four windows, each 3 ft. 6 in. wide?

522. At $2.75 a double roll, put on, what is the cost of papering a room 21 ft. long and 17 ft. 6 in. wide, if the room has two doors, each 4 ft. wide, and five windows, each 3 ft. 9 in. wide?

How many feet board measure in:

523. A board 16 ft. long, 9 in. wide, and 1 in. thick?

524. Three joists 12 ft. long, 6 in. wide, and 3 in. thick?

525. Twenty planks, 18 ft. long, 15 in. wide, and 4 in. thick?

526. A stick of timber 9 in. square, and 24 ft. long?

527. Ten boards, 15 ft. long, 14 in. wide, 1½ in. thick?

COMMON FRACTIONS.

528. 60 is $\frac{5}{12}$ of what number?

529. If a girl knits $\frac{3}{4}$ of a stocking in a day, how many days will it take her to knit 30 pairs of stockings?

530. A tailor cut $8\frac{2}{3}$ yards from a piece of cloth and had $17\frac{3}{4}$ yards left. How many yards had he at first?

531. If $3\frac{3}{4}$ yards of cloth are required for a shirt, how many shirts can be made from $56\frac{1}{4}$ yards of cloth?

532. What is the value of 5 pieces of cloth, each containing $26\frac{3}{4}$ yards, at \$1.25 a yard?

533. A train passes over $\frac{7}{12}$ of its route in $3\frac{1}{2}$ hours. In what time will it pass over its entire route?

534. Simplify $\frac{5}{8} - \frac{2}{3} + \frac{13}{24} - \frac{3}{16} + \frac{1}{6}$.

535. Change to decimals $\frac{3}{5}$; $\frac{7}{8}$; $\frac{9}{16}$; $\frac{1}{3}$; $\frac{7}{24}$; $\frac{119}{226}$.

536. Change to common fractions 0.875; 0.65; 0.385; 0.44.

537. How many minutes will it take at the rate of $2\frac{5}{6}$ gallons per minute to fill a cistern holding $61\frac{5}{6}$ gallons?

538. One boy does $\frac{2}{9}$, a second $\frac{1}{3}$, and a third $\frac{2}{7}$ of a piece of work. What fraction of the work remains to be done?

539. A merchant bought $38\frac{3}{16}$ pounds of butter and sold $16\frac{3}{4}$ pounds. How many pounds did he have left? .

540. What is the cost of 28 tons of coal at \$$5\frac{5}{8}$ per ton?

541. Find the sum and the difference of $\frac{5}{6}$ and $\frac{4}{9}$.

542. If $\frac{5}{6}$ of a ton of hay can be bought for \$$12\frac{1}{2}$, what part of a ton can be bought for \$3?

543. If a farmer can plough $1\frac{1}{5}$ acres of land in one day, how many days will it take him to plough $23\frac{2}{5}$ acres?

544. What number multiplied by $5\frac{7}{8}$ will make $18\frac{4}{5}$?

545. The product of three numbers is 14 and two of the numbers are $3\frac{5}{6}$ and $1\frac{5}{7}$. What is the third number?

546. A farm is composed of four fields that contain respectively $13\frac{3}{4}$ acres, $19\frac{4}{5}$ acres, $26\frac{2}{3}$ acres, and $32\frac{2}{15}$ acres. How many acres does the farm contain?

547. If a barrel of apples contains $2\frac{3}{4}$ bushels, how many barrels will $9\frac{5}{8}$ bushels of apples fill?

548. If 4 is added to each term of the fraction $\frac{7}{16}$, is its value increased or decreased, and by how much?

549. A horse and carriage were bought for $450 and the carriage cost $\frac{2}{7}$ as much as the horse. Find the cost of each.

550. If a man walks $22\frac{1}{2}$ miles a day, how many miles will he walk in $4\frac{4}{5}$ days?

551. A grocer sold $\frac{1}{4}$, then $\frac{2}{5}$, then $\frac{3}{10}$ of his sugar and had 56 pounds left. How many pounds had he at first?

552. How many times is $5\frac{4}{9}$ contained in 77?

553. If $19\frac{4}{5}$ pounds of sugar cost 99 cents, what will $32\frac{2}{7}$ pounds of sugar cost?

554. A worked $9\frac{1}{2}$ days, B $8\frac{3}{4}$ days, and C $8\frac{1}{2}$ days. How much did they all earn if each man earned $1\frac{1}{2}$ per day?

555. A bought 744 pounds of sugar at $4\frac{5}{8}$ cents a pound and sold it at $5\frac{1}{4}$ cents a pound. How much did he gain?

556. A farmer sold $\frac{2}{9}$ of a field, and then sold $\frac{2}{7}$ of the remainder. If he had $13\frac{1}{3}$ acres left, how many acres were there in the field?

557. A boy being asked to find the value of $10\frac{4}{15} + 2\frac{7}{9} + 3\frac{2}{3} + 4\frac{2}{3}$, gave as his answer 22. How great was his error?

PERCENTAGE.

558. Express 20%, 25%, 33⅓%, 87½%, 83⅓%, 45%, 6¼% as common fractions.

559. Express as a rate per cent ½; ⅓; ¾; ⅚; ⅗; $\frac{9}{10}$; $\frac{13}{20}$.

560. What is 37½% of 928 ?

561. What is 42% of $1946.50 ?

562. If gunpowder contains 75% of saltpetre, 10% of sulphur, and 15% of charcoal, how many pounds of each are there in a ton of gunpowder ?

563. A house worth $4000 rents for $360 a year. For what per cent of its value does the house rent ?

564. A farmer raised 275 bushels of potatoes one year and the next year 32% more. How many bushels of potatoes did he raise the second year ?

565. A farmer bought a cow for $30 and sold her for $36. What was his gain per cent ?

566. If rosin is melted with 20% of its weight of tallow, what per cent of tallow does the mixture contain ?

567. The population of the United States in 1880 was 50,155,783, and in 1890 was 62,622,250. What was the increase per cent ?

568. In a school there are 81 girls, and the number of girls is 45% of the whole number of pupils. How many pupils are there in the school ?

569. A town, after decreasing 25%, has 4539 inhabitants. Find its population at first.

570. What is the rate of commission when $225 is paid for selling $15,000 worth of goods ?

571. A man bought a house for $3000 and in selling it gained 25%. What was the selling price?

572. A man sold a house for $3000 and gained 20%. What was the cost of the house?

573. An agent in Chicago bought 1200 bushels of wheat at 72½ cents. How much was his commission at 1½%?

574. Find the net amount of a bill of $630.25, if a discount of 5% is allowed for cash.

575. A house worth $6000 is insured for ¾ its value at ¼ of 1% per annum. What is the annual premium?

576. A collector collected 85% of a debt of $1274, and charged 5% of the amount he collected. What was the net amount for the creditor?

Find the interest on:

577. $8263.45 for 7 yr. 3 mo. 18 dy. at 6%.

578. $6427.95 for 2 yr. 7 mo. 20 dy. at 6%.

579. $4283.52 for 4 yr. 9 mo. 15 dy. at 6%.

580. $3786.15 for 1 yr. 3 mo. 23 dy. at 6%.

581. $1215.17 for 3 yr. 2 mo. 12 dy. at 4½%.

582. $1862.45 for 2 yr. 11 mo. 24 dy. at 4%.

583. $2389.90 for 1 yr. 7 mo. 28 dy. at 5%.

584. $1475.45 for 4 yr. 5 mo. 17 dy. at 3½%.

585. Find the bank discount and the proceeds of a note for $1250, due in 6 mo., without grace, discounted at 5%.

586. Find the bank discount and the proceeds of a note for $2225, due in 4 mo., with grace, discounted at 4½%.

587. Find the bank discount and the proceeds of a note for $790, due in 90 days, with grace, discounted at 4%.

ANSWERS.

Lesson 35. Page 145.

1. 1102.	7. 1447.	13. 133.77.	19. 438,997.
2. 1445.	8. 14,219.	14. 222,038.	20. 527.4053.
3. 1982.	9. 18,678.	15. 209,381.	21. 171.8762.
4. 986.	10. 26,405.	16. 260.164.	22. 200.8964.
5. 1712.	11. 25,934.	17. 72.4347.	23. 35.351.
6. 1283.	12. 231.84.	18. $636.43.	24. 1491.4375.
			25. $194.36.

Lesson 36. Page 146.

1. $18,214.	3. 2,480,195.	5. 1,602,778.	7. 1,019,067.
2. 2017.	4. 1,638,162.	6. 1,471,784.	8. 711,998.

Lesson 37. Page 147.

1. 704.	6. 1921.	11. 11,878.	16. 116,689.
2. 381.	7. 3868.	12. 2795.	17. 457,547.
3. 523.	8. 350.	13. 10,748.	18. 41,799.
4. 426.	9. 2529.	14. 8757.	19. 85,216.
5. 159.	10. 1316.	15. 31,407.	20. 24,184.

Lesson 38. Page 148.

1. 0.06.	9. 5.855.	17. 18.7132.	24. 33.151.
2. 0.78.	10. 2.759.	18. 5.8908.	25. 0.0023.
3. 0.893.	11. 0.668.	19. 1.0276.	26. 747.8268.
4. 2.306.	12. 1.857.	20. 0.9558.	27. 761.613.
5. 0.067.	13. 0.885.	21. 2.475.	28. 18.777.
6. 0.107.	14. 0.072.	22. 74.2425.	29. 57.6246.
7. 2.224.	15. 0.505.	23. 0.5176.	30. 5.8435.
8. 0.882.	16. 3.1989.		

Lesson 39. Page 149.

1. 52.	4. 1106.	7. 71,041.	10. 51.
2. 66.	5. 2920.	8. 109,008.	11. $1.17.
3. 1782.	6. 76,831.	9. 14,095.	

Lesson 40. Page 150.

1. 7374.	5. 9765.	9. 34,260.	13. 26,394.
2. 9566.	6. 14,245.	10. 8256.	14. 16,394.
3. 8637.	7. 14,263.	11. 33,882.	15. 34,584.
4. 10,971.	8. 10,344.	12. 18,744.	16. 46,935.

17. 31,304.
18. 41,335.
19. 40,524.
20. 54,978.
21. 65,688.
22. 34,696.
23. 29,355.
24. 26,082.
25. 16,338.

26. 15,138.
27. 49,445.
28. 35,908.
29. 58,668.
30. 27,153.
31. 30,195.
32. 53,400.
33. 60,501.
34. 61,686.

35. 117,416.
36. 352,640.
37. 340,278.
38. 220,969.
39. 300,656.
40. 504,126.
41. 312,741.
42. 332,343.
43. 658,674.

44. 206,712.
45. 688,275.
46. 508,624.
47. 607,401.
48. 270,879.
49. 438,921.
50. 399,063.
51. 793,384.

Lesson 41. Page 151.

1. 3648.
2. 8512.
3. 20,440.
4. 8556.
5. 18,112.
6. 26,508.
7. 14,763.
8. 20,444.

9. 27,553.
10. 69,184.
11. 34,272.
12. 56,066.
13. 72,412.
14. 47,058.
15. 62,568.
16. 3,566,541.

17. 4,235,374.
18. 5,952,816.
19. 5,921,580.
20. 4,212,032.
21. 1,601,613.
22. 786,714.
23. 4,533,573.
24. 2,722,225.

25. 4,175,712.
26. 6,418,652.
27. 3,412,836.
28. 5,356,521.
29. 3,276,303.
30. 6,731,472.

Lesson 42. Page 152.

. 4670.
. 31,200.
. 587,000.
. 18,336,000.
. 29,124,000.
. 40,635,000.
1. 86,140,000.

8. 101,088,000.
9. 194,880,000.
10. 350,420,000.
11. 104,832,000.
12. 97,290,000.
13. 50,430,400.
14. 49,854,240.

15. 11,428,368,000.
16. 172,437,740,000.
17. 10,800.
18. 48,000.
19. 108,000.
20. $210.

Lesson 43. Page 153.

1. 240.204.
2. 197.896.
3. 1769.08.
4. 55.5676.
5. 367,848.
6. 232.379.

7. 6601.68.
8. 5165.71.
9. 433.5.
10. 0.8421.
11. 34.704.
12. 0.0164.

13. 11.9385.
14. 91.008.
15. 101.1725.
16. 21015.984.
17. 3417.
18. 9550.

19. 6828.467.
20. 54.2913.
21. 25.6275.
22. 87.0672.
23. 4603.8601.
24. 4954.6497.

Lesson 44. Page 154.

1. 109,500.
2. 57,096.
3. $67.20.

4. $97.75.
5. $66.50.
6. $999.

7. 1140.
8. $567.
9. 354.36.

10. 14,560 ft.
11. $32.40.

Lesson 48. Page 158.

1. 217.
2. 292.
3. 149.
4. 108.

5. 36.
6. 143.
7. 175.
8. 103.

9. 54.
10. 117.
11. 121.
12. 115.

13. 113.
14. 105.
15. 69.
16. 256—1.

17. 235–2.
18. 211–1.
19. 180–1.
20. 143–4.
21. 124–4.
22. 100–7.
23. 2897.
24. 1958.
25. 1424.
26. 1795.
27. 559.
28. 168.
29. 1071.

30. 327.
31. 1855–1.
32. 1075–1.
33. 2116–3.
34. 1914–3.
35. 1163–5.
36. 1237.
37. 541–1.
38. 917–3.
39. 1959–2.
40. 817–8.
41. 817–3.
42. 770–6.

43. 3250–1.
44. 979–4.
45. 47,937.
46. 15,291.
47. 11,593.
48. 15,659.
49. 11,062.
50. 13,226.
51. 10,978.
52. 10,908.
53. 11,279–3.
54. 8247–2.
55. 12,766–2.

56. 27,873–2.
57. 14,353–5.
58. 4290–2.
59. 1769–3.
60. 6260.
61. 9341.
62. 15,085–4.
63. 5214–1.
64. 3279–3.
65. 8173–3.
66. 19,049–1.

Lesson 50. Page 160.

1. 164–34.
2. 155–8.
3. 201–4.
4. 141–37.
5. 149–36.
6. 88–66.
7. 109–26.
8. 170–9.
9. 218–24.

10. 109–11.
11. 126–14.
12. 128–52.
13. 130–47.
14. 101–21.
15. 126–14.
16. 129–23.
17. 118–27.
18. 105–17.

19. 117–38.
20. 113–30.
21. 138–5.
22. 141–8.
23. 222–9.
24. 171–21.
25. 117–25.
26. 106–30.
27. 165–17.

28. 154–5.
29. 317–20.
30. 159–33.
31. 606–32.
32. 428–95.
33. 127–258.
34. 116–36.
35. 24–338.
36. 138–2.

Lesson 51. Page 161.

1. 139–389.
2. 278–54.
3. 145–162.
4. 129–157.
5. 109–196.
6. 122–290.
7. 134–578.
8. 79–164.
9. 53–159.
10. 227–129.
11. 237–210.
12. 108–420.
13. 121–135.
14. 103–622.

15. 112–618.
16. 127–336.
17. 195–80.
18. 106–113.
19. 219–207.
20. 211–200.
21. 141–318.
22. 108–825.
23. 97–465.
24. 147–540.
25. 1211–427.
26. 1160–105.
27. 807–12.
28. 223–805.

29. 1024–494.
30. 761–173.
31. 1363–134.
32. 830–610.
33. 682–69.
34. 884–110.
35. 670–526.
36. 724–80.
37. 2315–55.
38. 1347–189.
39. 1009–210.
40. 1774–323.
41. 654–152.
42. 2419–285.

43. 1298–187.
44. 1182–273.
45. 4740–184.
46. 153–3330.
47. 406–1106.
48. 126–486.
49. 125–3932.
50. 140–3958.
51. 108–4761.
52. 127–464.
53. 83–2717.
54. 148–1854.

Lesson 52. Page 162.

1. $12.
2. 8 days.
3. $35.

4. 36 cents.
5. $48.
6. 9.
7. 55 cents.

8. $56.
9. 5.
10. $14.
11. $20.

12. 56 cents.
13. 12 days.
14. 6 pounds.

Lesson 1. Page 163.

1. 1.09.	6. 2.31.	11. 22.1.	16. 33.8.
2. 1.16.	7. 3.13.	12. 47.3.	17. 0.131.
3. 1.15.	8. 2.8.	13. 2.34.	18. 1.64.
4. 2.32.	9. 1.12.	14. 0.653.	19. 1.21.
5. 3.11.	10. 22.4.	15. 3.72.	20. 1.22.
			21. 23.1.

Lesson 2. Page 164.

1. 430.	11. 50.	21. 90.	31. 3.2.
2. 305.	12. 60.	22. 43.	32. 29.
3. 272.	13. 60.	23. 31.	33. 2.6.
4. 290.	14. 90.	24. 27.	34. 4.8.
5. 230.	15. 1100.	25. 3.1.	35. 1.1.
6. 160.	16. 1100.	26. 2.3.	36. 2.2.
7. 130.	17. 1300.	27. 16.	37. 22.
8. 400.	18. 140.	28. 3.6.	38. 310.
9. 250.	19. 1600.	29. 44.	39. 140.
10. 402.	20. 180.	30. 4.02.	

Lesson 3. Page 165.

1. 10.03.	9. 0.017.	17. 20,000.	25. 35,900.
2. 3.1416.	10. 7.8.	18. 500.	26. 24,163,000.
3. 5.4.	11. 6.48.	19. 1200.	27. 7.46.
4. 8.17.	12. 2100.	20. 2480.	28. 0.04.
5. 115.1875.	13. 130.	21. 20.3.	29. 40.
6. 3692.	14. 5025.	22. 8.302.	30. 400.
7. 0.312.	15. 1040.	23. 0.672.	31. 4900.
8. 88.	16. 3,209,000.	24. 240.6.	32. 0.04.

Lesson 4. Page 166.

1. 118.	. 170.	8. 256.	12. 5.5 cts.
2. $7.50.	29.	9. $125.	13. 28.
3. $2003.	. 17.	10. 109.	14. 24.
	4. 18.	11. 6.25.	

Lesson 5. Page 167.

1. 1.2109.	3. 24.2985.	6. 0.0029.	9. 0.0008.
2. 3.1913.	4. 140.6923.	7. 0.0136.	10. 0.0001.
	5. 0.0082.	8. 0.0133.	

Lesson 6. Page 168.

1. 5,798,758 tons.	4. $693,048,702.	7. 41.64 bu.
2. 14.28 times.	5. $69,786,800.	8. 41 bu.
3. 1,295,179 tons.	6. 165,831 acres.	9. 14 bu.

Lesson 8. Page 170.

1. 13 pt.	3. 57 pt.	5. 65 pt.	7. 252 qt.
2. 7 pt.	4. 36 gi.	6. 90 pt.	8. 1512 pt.

9. 28 gal. 2 qt. 1 pt.
10. 6 gal. 1 qt. 1 pt.

11. 45 gal. 2 qt. 1 pt.
12. 27 gal. 3 qt.

13. 131 gal. 2 qt.
14. 53 gal. 3 qt.
   1 pt. 3 gi.

Lesson 9. Page 171.

1. 20 gal. 0 qt. 1 pt.
2. 48 gal. 1 qt.

3. 103 gal. 1½ pt.
4. 13 gal. 3 qt.

5. 10 gal. 3 qt. 1 pt.
6. 9 gal. 3 qt. ½ pt.
7. 21 gal. 1 qt. 1 pt.

Lesson 10. Page 172.

1. 70 gal. 3 qt. 1 pt.
2. 220 gal. 2 qt.

3. 31 qt.
4. 21 gal. 1⅙ pt.

5. 15 gal. 3 qt.
6. 16 gal. 1 qt. 1 pt.

Lesson 11. Page 173.

1. 188 qt.
2. 63 bu. 1 pk. 4 qt.
3. 69 bu. 1 pk. 7 qt.
4. 3 bu. 2 pk. 6 qt.

5. 21.
6. 2 bu. 2 pk. 2 qt.
7. 16 bu. 3 pk. 5 qt.
8. 28 bu. 2 pk. 6 qt.

9. 411 bu. 1 pk. 2 qt.
10. 2 bu. 3 qt.
11. 3 bu. 1 pk. 7 qt.
12. 13 bu. 2 pk. 6 qt.

Lesson 12. Page 174.

1. 8174 lb.
2. 39 t. 596 lb.

3. 18 t.
4. 1 t. 500 lb.

5. 5 t. 614 lb.
6. 1 t. 500 lb.

7. 28 lb. 6 oz.
8. 81 cents.
9. $56.25.

Lesson 13. Page 175.

1. 1240.
2. 172 dwt.
3. 7 lb. 4 oz.
4. 480.

5. 6 oz. 15 dwt.
6. 80 cents.
7. 24.
8. 109.

9. 53 oz. 6 dwt.
10. 165 oz. 18 dwt.
11. 29 oz 18 dwt. 3 gr.
12. 4 oz. 0 dwt.

Lesson 14. Page 176.

1. 5012 min.
2. 27,050 sec.
3. 14 dy. 4 hr.

4. 5 dy. 13 hr. 40 min.
5. 1 yr. 2 dy. 1 hr.
6. 1 wk. 1 dy. 9 hr.

7. 18 dy. 10 hr.
8. 5 dy. 16 hr. 18 min. 3 sec.
9. 4 dy. 10 hr. 42 min.

Lesson 16. Page 178.

1. 211 in.
2. 100 in.
3. 908 rd.
4. 5301 rd.
5. 3 mi.
6. 170 rd.
7. 2 mi. 80 rd.

8. 11 mi.
9. 48 yd. 11 in.
10. 58 yd. 1 ft. 5 in.
11. 30 mi. 89 rd. 4½ yd.
12. 39 mi. 275 rd. 14 ft.
13. 18 mi. 262 rd. 5 ft.
14. 1 mi. 272 rd. 15 ft. 11 in.

15. 10 yd. 2 ft. 11 in.
16. 18 yd. 6 in.
17. 6 mi. 157 rd. 2½ ft.
18. 8 mi. 187 rd. 3½ ft.
19. 3 mi. 64 rd. 4½ yd.
20. 9 mi. 32 rd. 4 yd.
21. 171 yd. 1 ft. 3 in.
22. 283 rd. 5 yd. 2 ft.

Lesson 17. Page 179.

1. 60 ft.
2. 84 ft.

3. 68 ft.
4. 90 ft.

5. 66 in. ; 88 in. ; 22 ft.
6. 3.5 in.
7. 10.5 in.

Lesson 18. Page 180.

1. 4570 sq. ft.   4. 312 sq. ft. 68 sq. in.   7. 13 A. 75 sq. rd.
2. 5 A.   5. 14 A.   8. 9 sq. yd. 7 sq. ft. 20 sq. in.
3. 570 sq. rd.   6. 533 A. 36 sq. rd.   9. 19 A. 20 sq. rd.

Lesson 19. Page 181.

1. 40 sq. in.   5. 90 sq. in.   9. 16 sq. ft.   13. 4 sq. yd.
2. 54 sq. in.   6. 64 sq. in.   10. 420 sq. ft.   14. 986 sq. yd.
3. 56 sq. in.   7. 6 sq. ft.   11. 270 sq. in.   15. 80 sq. rd.
4. 110 sq. in.   8. 8 sq. ft.   12. 36 sq. ft.   16. 594 sq. ft.

Lesson 20. Page 182.

1. 860 sq. ft.   3. 72 sq. yd.   5. 16 sq. yd.   7. 112 sq. in.
2. 30 sq. yd.   4. 10 A.   6. 32 sq. yd.   8. 122 sq. in.
10. 314 sq. in.; 805 sq. in.; 1257 sq. in.   9. 54 sq. ft.
11. 805 sq. in.; 1018 sq. in.
12. 380 sq. in.; 616 sq. in.
13. 707 sq. in.; 1257 sq. in.; 1521 sq. in.; 1663 sq. in.
14. 1964 sq. in.

Lesson 21. Page 183.

1. 5 strips.   2. 10 strips.   3. 35 yd.; $\frac{1}{3}$ yd.

Lesson 22. Page 184.

1. 8.   2. $8.00.   3. $4.50.   4. 10.   5. 9.

Lesson 23. Page 185.

1. 16 ft.   4. $16\frac{2}{3}$ ft.   7. 90 ft.   10. $133\frac{1}{3}$ ft.
2. 24 ft.   5. 12 ft.   8. 32 ft.   11. 16 ft.
3. 14 ft.   6. 15 ft.   9. 40 ft.   12. 35 ft.

Lesson 24. Page 186.

1. 372 cu. ft.   4. 175 cu. yd. 7 cu. ft.   7. 10 cd.
2. 22 cu. yd. 6 cu. ft.   5. 1 cu. yd. 16 cu. ft.   8. 15 cd. 96 cd. ft.
3. 23 cu. yd. 25 cu. ft.   6. 64 cu. yd. 25 cu. ft.   9. 4 cu. yd. 20 cu. ft.
10. $9.

Lesson 25. Page 187.

1. 96 cu. in.   4. 84 cu. in.   7. 30 cu. ft.
2. 48 cu. in.   5. 320 cu. in.   8. 462 cu. yd.
3. 64 cu. in.   6. 27 cu. in.   9. 264 cu. yd.

Lesson 28. Page 190.

1. $9.46.   2. $21.45.   3. $24.88.   4. $18.24.

Lesson 33. Page 195.

1. $\frac{18}{7}$.   3. $\frac{56}{7}$.   5. $\frac{99}{9}$.   7. $\frac{156}{12}$.   9. $\frac{405}{15}$.   11. $\frac{325}{25}$.
2. $\frac{56}{8}$.   4. $\frac{54}{6}$.   6. $\frac{132}{11}$.   8. $\frac{247}{13}$.   10. $\frac{240}{20}$.   12. $\frac{700}{50}$.

Lesson 34. Page 196.

1. $\frac{23}{2}$.   9. $\frac{287}{100}$.   17. $\frac{234}{11}$.   25. $\frac{211}{12}$.   33. $\frac{97}{14}$.
2. $\frac{51}{4}$.   10. $\frac{172}{13}$.   18. $\frac{125}{7}$.   26. $\frac{406}{25}$.   34. $\frac{107}{12}$.
3. $\frac{71}{12}$.   11. $\frac{281}{50}$.   19. $\frac{121}{9}$.   27. $\frac{759}{50}$.   35. $\frac{96}{13}$.
4. $\frac{77}{16}$.   12. $\frac{175}{36}$.   20. $\frac{137}{11}$.   28. $\frac{289}{20}$.   36. $\frac{142}{15}$.
5. $\frac{204}{25}$.   13. $\frac{383}{19}$.   21. $\frac{181}{15}$.   29. $\frac{273}{20}$.   37. $\frac{147}{50}$.
6. $\frac{95}{6}$.   14. $\frac{175}{17}$.   22. $\frac{191}{10}$.   30. $\frac{296}{25}$.   38. $\frac{337}{100}$.
7. $\frac{118}{7}$.   15. $\frac{129}{5}$.   23. $\frac{205}{11}$.   31. $\frac{103}{27}$.   39. $\frac{591}{100}$.
8. $\frac{207}{8}$.   16. $\frac{164}{9}$.   24. $\frac{385}{36}$.   32. $\frac{83}{17}$.   40. $\frac{419}{50}$.

Lesson 35. Page 197.

1. 3.   5. 7.   9. $3\frac{3}{4}$.   13. $17\frac{1}{2}$.   17. $6\frac{3}{4}$.   21. $3\frac{9}{14}$.
2. 6.   6. 9.   10. $6\frac{3}{4}$.   14. $3\frac{5}{6}$.   18. $3\frac{1}{3}$.   22. $3\frac{1}{12}$.
3. 2.   7. $3\frac{3}{5}$.   11. $6\frac{2}{9}$.   15. $3\frac{1}{8}$.   19. $4\frac{3}{5}$.   23. $2\frac{1}{36}$.
4. 6.   8. $9\frac{2}{3}$.   12. $5\frac{8}{11}$.   16. $4\frac{3}{5}$.   20. $3\frac{1}{3}$.   24. $5\frac{11}{16}$.

1. $\frac{2}{3}$.   5. $\frac{3}{7}$.   9. $\frac{3}{4}$.   13. $\frac{2}{9}$.   17. $\frac{1}{2}$.   21. $\frac{3}{4}$.
2. $\frac{2}{3}$.   6. $\frac{1}{8}$.   10. $\frac{4}{5}$.   14. $\frac{5}{8}$.   18. $\frac{3}{4}$.   22. $\frac{2}{3}$.
3. $\frac{4}{5}$.   7. $\frac{7}{9}$.   11. $\frac{4}{5}$.   15. $\frac{1}{3}$.   19. $\frac{2}{3}$.   23. $\frac{3}{8}$.
4. $\frac{1}{2}$.   8. $\frac{3}{5}$.   12. $\frac{2}{5}$.   16. $\frac{1}{4}$.   20. $\frac{3}{5}$.   24. $\frac{1}{4}$.

Lesson 36. Page 198.

1. $\frac{3}{7}$.   4. $\frac{2}{7}$.   7. $\frac{2}{5}$.   10. $\frac{2}{11}$.   13. $\frac{5}{9}$.   16. $\frac{1}{3}$.
2. $\frac{4}{5}$.   5. $\frac{3}{11}$.   8. $\frac{3}{13}$.   11. $\frac{3}{16}$.   14. $\frac{1}{6}$.   17. $\frac{2}{3}$.
3. $\frac{5}{11}$.   6. $\frac{5}{16}$.   9. $\frac{4}{20}$.   12. $\frac{4}{21}$.   15. $\frac{4}{3}$.   18. $\frac{15}{5}$.

Lesson 37. Page 199.

1. 5.   4. 10.   7. $9\frac{1}{2}$.   10. $8\frac{1}{2}$.   13. $16\frac{4}{5}$.   16. 12.   19. $5\frac{3}{5}$.
2. 7.   5. $6\frac{1}{2}$.   8. $9\frac{1}{2}$.   11. 14.   14. $12\frac{2}{9}$.   17. 18.   20. $10\frac{5}{6}$.
3. $6\frac{2}{3}$.   6. $15\frac{3}{5}$.   9. 13.   12. $16\frac{1}{4}$.   15. 11.   18. 24.   21. 16.

1. 15.   3. 20.   5. 25.   7. 22.   9. 70.   11. $150\frac{1}{2}$.   13. 17.   15. 22.
2. 14.   4. 21.   6. 30.   8. 24.   10. 40.   12. $184\frac{1}{4}$.   14. $27\frac{1}{2}$.

Lesson 38. Page 200.

1. 2.
2. $\frac{1}{2}$.
3. $\frac{5}{6}$.
4. $1\frac{1}{5}$.
5. $1\frac{1}{9}$.
6. $\frac{2}{5}$.
7. $2\frac{1}{2}$.
8. 15.
9. $\frac{1}{15}$.
10. $1\frac{1}{3}$.
11. $1\frac{1}{4}$.
12. $\frac{4}{5}$.
13. $\frac{4}{5}$.
14. $\frac{4}{7}$.
15. $1\frac{2}{3}$.
16. $1\frac{1}{6}$.
17. $\frac{3}{4}$.
18. $\frac{3}{4}$.

$\frac{3}{5}$. $1\frac{1}{4}$.   . 4.
19. $1\frac{7}{9}$.
23. 3.
24. 3.

25. $1\frac{1}{3}$.
26. 2.
27. $\frac{1}{2}$.
28. $\frac{3}{4}$.
29. 2.
30. $\frac{1}{2}$.
31. 5.
32. 3.
33. $1\frac{1}{3}$.
34. 2.
35. $\frac{1}{2}$.
36. 5.

Lesson 39. Page 201.

1. $\frac{4}{12}, \frac{9}{12}, \frac{10}{12}$.
2. $\frac{3}{12}, \frac{2}{12}, \frac{1}{12}$.
3. $\frac{9}{18}, \frac{4}{18}, \frac{15}{18}$.
4. $\frac{4}{8}, \frac{6}{8}, \frac{5}{8}$.
5. $\frac{6}{18}, \frac{10}{18}, \frac{7}{18}$.
6. $\frac{16}{24}, \frac{9}{24}, \frac{5}{24}$.
7. $\frac{7}{14}, \frac{10}{14}, \frac{3}{14}$.
8. $\frac{7}{21}, \frac{9}{21}, \frac{4}{21}$.
9. $\frac{3}{15}, \frac{5}{15}, \frac{1}{15}$.
10. $\frac{14}{42}, \frac{30}{42}, \frac{27}{42}$.
11. $\frac{15}{24}, \frac{20}{24}, \frac{5}{24}$.
12. $\frac{21}{28}, \frac{11}{28}, \frac{8}{28}$.
13. $\frac{15}{36}, \frac{7}{36}, \frac{20}{36}$.
14. $\frac{12}{42}, \frac{9}{42}, \frac{5}{42}$.
15. $\frac{45}{75}, \frac{35}{75}, \frac{24}{75}$.

Lesson 42. Page 204.

1. $1\frac{7}{15}$.
2. $1\frac{7}{12}$.
3. $1\frac{19}{30}$.
4. $1\frac{1}{4}$.
5. $1\frac{5}{18}$.
6. $1\frac{2}{5}$.
7. $1\frac{7}{24}$.
8. $1\frac{4}{9}$.
9. $1\frac{2}{3}$.
10. $1\frac{1}{10}$.
11. $1\frac{5}{42}$.
12. $1\frac{1}{6}$.
13. $1\frac{19}{30}$.
14. $\frac{7}{12}$.
15. $2\frac{23}{60}$.
16. $1\frac{17}{18}$.
17. $2\frac{5}{24}$.
18. $1\frac{2}{5}$.
19. $1\frac{29}{30}$.
20. $1\frac{27}{28}$.
21. $2\frac{4}{9}$.

1. $8\frac{4}{9}$.
2. $9\frac{5}{8}$.
3. $13\frac{1}{16}$.
4. $14\frac{3}{20}$.
5. $17\frac{1}{6}$.
6. $15\frac{5}{18}$.
7. $14\frac{1}{4}$.
8. $22\frac{1}{12}$.
9. $20\frac{17}{20}$.
10. $18\frac{5}{8}$.
11. $16\frac{9}{16}$.
12. $23\frac{29}{30}$.
13. $21\frac{5}{8}$.
14. $22\frac{3}{14}$.

Lesson 43. Page 205.

1. $\frac{5}{21}$.
2. $\frac{1}{8}$.
3. $\frac{7}{15}$.
4. $\frac{7}{24}$.
5. $\frac{1}{12}$.
6. $\frac{9}{16}$.
7. $\frac{3}{20}$.
8. $\frac{4}{15}$.
9. $\frac{5}{12}$.
10. $\frac{4}{17}$.
11. $\frac{14}{15}$.
12. $\frac{1}{8}$.
13. $\frac{1}{4}$.
14. $\frac{7}{12}$.
15. $\frac{5}{21}$.

1. $11\frac{1}{4}$.
2. $8\frac{1}{12}$.
3. $8\frac{1}{12}$.
4. $8\frac{1}{6}$.
5. $13\frac{5}{6}$.
6. $13\frac{11}{12}$.
7. $51\frac{14}{15}$.
8. $41\frac{2}{3}$.
9. $2\frac{1}{6}$.
10. $1\frac{5}{6}$.
11. $43\frac{1}{18}$.
12. $42\frac{17}{18}$.
13. $39\frac{11}{12}$.
14. $40\frac{1}{12}$.
15. $19\frac{15}{16}$.
16. $20\frac{1}{16}$.
17. $22\frac{7}{12}$.
18. $25\frac{1}{10}$.
19. $1\frac{23}{40}$.
20. $2\frac{11}{20}$.
21. $1\frac{7}{8}$.
22. $1\frac{1}{24}$.
23. $2\frac{1}{3}$.
24. $3\frac{4}{63}$.

Lesson 44. Page 206.

1. $1\frac{2}{3}$.
2. $4\frac{4}{7}$.
3. $1\frac{1}{4}$.
4. $1\frac{1}{35}$.
5. $2\frac{7}{10}$.
6. $1\frac{2}{3}$.
7. 1.
8. $2\frac{17}{24}$.
9. $11\frac{11}{12}$.
10. $19\frac{5}{6}$.
11. $43\frac{13}{16}$.
12. $\frac{35}{36}$.
13. $\frac{5}{42}$.
14. $15\frac{5}{8}$.

Lesson 45. Page 207.

1. $\frac{10}{21}$.
2. $\frac{1}{4}$.
3. $\frac{16}{21}$.
4. $\frac{5}{32}$.
5. $10\frac{2}{3}$.
6. $\frac{9}{10}$.
7. $\frac{2}{45}$.
8. 2.
9. $2\frac{2}{3}$.
10. $1\frac{11}{52}$.
11. $\frac{91}{100}$.
12. $\frac{36}{49}$.
13. $1\frac{7}{25}$.
14. $2\frac{1}{4}$.
15. $1\frac{3}{5}$.
16. 16.
17. $9\frac{3}{5}$.
18. $3\frac{4}{5}$.
19. $2\frac{2}{3}$.
20. $9\frac{6}{11}$.
21. $77\frac{7}{9}$.

Lesson 46. Page 208.

1. $\frac{1}{4}$.
2. $\frac{2}{3}$.
3. $\frac{7}{8}$.
4. $\frac{2}{4}$.
5. $\frac{1}{4}$.
6. $\frac{2}{3}$.
7. $\frac{8}{9}$.
8. $\frac{2}{3}$.
9. $\frac{3}{4}$.
10. $\frac{3}{10}$.
11. $\frac{1}{4}$.
12. $\frac{4}{7}$.
13. $\frac{4}{5}$.
14. $\frac{5}{18}$.
15. $\frac{1}{4}$.
16. $\frac{8}{15}$.
17. $\frac{7}{6}$.
18. $\frac{1}{7}$.
19. $\frac{4}{15}$.
20. $\frac{2}{9}$.
21. $\frac{1}{12}$.
22. $\frac{1}{8}$.
23. $\frac{1}{6}$.
24. $\frac{1}{16}$.
25. $\frac{3}{8}$.
26. $\frac{1}{3}$.
27. $\frac{2}{3}$.
28. $\frac{5}{8}$.
29. $\frac{7}{8}$.
30. $\frac{5}{6}$.

Lesson 47. Page 209.

1. 24.
2. 35.
3. 24.
4. 78.
5. 48.
6. 112.
7. $9\frac{3}{5}$.
8. $12\frac{3}{5}$.
9. $11\frac{2}{3}$.
10. $12\frac{4}{5}$.
11. \$40; \$110; \$185.
12. \$97$\frac{1}{2}$.
13. 36 acres.
14. $6\frac{1}{4}$ barrels.
15. 50 bushels.
16. 143 miles.

Lesson 48. Page 210.

1. $\frac{2}{25}$.
2. $\frac{5}{8}$.
3. $\frac{4}{125}$.
4. $\frac{3}{8}$.
5. $\frac{1}{250}$.
6. $\frac{32}{125}$.
7. $\frac{17}{40}$.
8. $\frac{3}{200}$.
9. $7\frac{3}{40}$.
10. $3\frac{1}{8}$.
11. $1\frac{29}{40}$.
12. $7\frac{7}{8}$.
13. 0.06.
14. 0.15.
15. 0.025.
16. 0.04.
17. 0.135.
18. 0.032.
19. 0.004.
20. 17.875.
21. 5.375.
22. 7.075.
23. 1.9375.
24. 5.0625.

Lesson 49. Page 211.

1. 27 in.
2. 10 oz.
3. 1000 lb.
4. 24 cu. ft.
5. 120 sq. rd.
6. 112 cd. ft.
7. $1\frac{1}{2}$ pt.
8. 16 hr.
9. 45 min.
10. 8 qr.
11. 27 qr.; 29 qr.
12. 9 halves.
13. 11.
14. 2 gal.
15. 36 ct.
16. 21 ct.
17. 7 mi.
18. $7\frac{1}{2}$ mi.
19. $10\frac{1}{2}$ mi.
20. $\frac{5}{8}$ yd.
21. \$5$\frac{1}{4}$.
22. \$5$\frac{1}{4}$.
23. \$1.80.

Lesson 50. Page 212.

1. \$0.90.
2. \$0.81.
3. \$1.63.
4. \$1.78.
5. \$5.21.
6. \$1.98.
7. \$6.00.
8. \$27.00.
9. \$31.50.
10. \$0.95.
11. \$1.04.
12. \$0.84.
13. \$3.05.
14. \$2.02.
15. \$6.21.
16. \$29.48.
17. \$13.44.
18. \$19.25.
19. \$6.16.
20. \$2.24.
21. \$7.16.
22. \$15.60.
23. \$1.40.
24. \$506.25.
25. \$884.25.

Lesson 51.　Page 213.

1. $\frac{1}{3}$.
2. $\frac{5}{12}$.
3. $4\frac{1}{2}$.
4. $2.75.
5. 10 pounds.

6. $1\frac{3}{4}$ pecks.
7. $\frac{19}{3}$; $\frac{27}{5}$; $\frac{59}{9}$; $\frac{60}{7}$; $\frac{38}{3}$.
8. $9\frac{1}{2}$; $3\frac{4}{7}$; $3\frac{5}{8}$; $2\frac{9}{11}$; $3\frac{5}{12}$; $10\frac{8}{9}$.
9. $\frac{4}{5}$; $\frac{2}{4}$; $\frac{7}{9}$; $\frac{2}{3}$; $\frac{2}{3}$.
10. 15 bushels.

11. $1.08.
12. $28\frac{1}{8}$.
13. 5 cows.
14. 96 barrels.
15. 75 cents.

Lesson 52.　Page 214.

1. $56.
2. $\frac{1}{2}$ ton.
3. $5\frac{1}{4}$ pounds.
4. 7 yards.
5. $21\frac{3}{5}$ miles.

6. $\frac{1}{2}$.
7. 9 barrels.
8. $\frac{8}{8}$.
9. 40 marbles.
10. $250.

11. $150.
12. $\frac{1}{3}$.
13. $364.
14. $\frac{1}{2}$; $\frac{1}{4}$.

Lesson 53.　Page 215.

1. $180\frac{1}{7}$.
2. 1.
3. $69\frac{1}{3}$.
4. $1\frac{2}{3}$.

5. $\frac{1}{5}$.
6. $24\frac{4}{8}$.
7. 10 days.
8. $2100.

9. 1.725.
10. $7\frac{7}{8}$.
11. $\frac{5}{8}$.
12. $75.

13. 36 yards.
14. $\frac{8}{21}$.
15. 14 yards.
16. $\frac{9}{20}$.

Lesson 54.　Page 216.

1. 1920.
2. $3250.
3. $43\frac{1}{2}$.

4. $8.10.
5. 192 acres.
6. $469\frac{1}{3}$.

7. $13\frac{1}{8}$.
8. $\frac{7}{8}$.
9. $15\frac{5}{4}$ gallons.

10. $2\frac{17}{18}$.
11. $3\frac{16}{25}$ hours.
12. $15\frac{9}{113}$ feet.

Lesson 56.　Page 218.

1. 10.5.
2. 75.

3. 80 men.
4. 36 sheep.

5. 156 dy.
6. $280.

7. 608 oz.
8. 1012 lb.

9. 1530 ft.
10. 206 mi.

1. $33\frac{1}{3}\%$.
2. $2\frac{1}{2}\%$.

3. 300%.
4. 4000%.

5. 20%.
6. 20%.

7. 75%.
8. $12\frac{8}{11}\%$.

9. $89\frac{2}{9}\%$.
10. $82\frac{2}{7}\%$.

Lesson 57.　Page 219.

1. 28.
2. 40.
3. 120.

4. 80.
5. 90.
6. 54.

7. 24.
8. 600.
9. 800.

10. 300.
11. 200.
12. 400.

13. 600.
14. 160.

Lesson 58.　Page 220.

1. $1600.
2. 15.
3. $2.40.

4. $2000.
5. 10%.
6. $120.

7. $200.
8. 8%.
9. 500.

10. 1232 pounds.
11. $146.
12. $0.80.

Lesson 59.　Page 221.

1. $100.
2. $100.

3. $8120.
4. $4000.

5. 2%.
6. $15,200.

7. $2\frac{1}{2}\%$.
8. $4.16\frac{2}{3}$.

9. 150 bushels.
10. $13,960.

Lesson 60. Page 222.

1. $120.
2. $90.
3. $180.
4. $13,000.
5. $4800.
6. $3060.
7. $7900.
8. $67.50.
9. $\frac{1}{4}$%.
10. $\frac{1}{2}$%.
11. $500; $800; $700.

Lesson 61. Page 223.

1. $125.
2. $41.
3. $240.
4. $25.
5. $50.
6. $12.50.
7. $1.50.
8. $20.
9. $45.
10. $34.
11. $36.
12. $1235.
13. $61.50.
14. $1000.
15. $0.22.

Lesson 62. Page 224.

1. $18.31.
2. $41.88.
3. $7.82.
4. $103.14.
5. $131.40.
6. $106.27.
7. $316.99.
8. $20.
9. $30.
10. $1.26.

Lesson 63. Page 225.

1. $18.
2. $24.75.
3. $124.
4. $82.29.
5. $19.88.
6. $32.81.
7. $0.11.
8. $1.76.
9. $6.61.
10. $765.63.
11. $673.61.
12. $532.59.
13. $557.96.
14. $615.82.
15. $816.97.
16. $731.87.
17. $507.80.
18. $822.97.

Lesson 64. Page 226.

1. $44.87.
2. $32.05.
3. $43.27.
4. $28.85.
5. $32.05.
6. $29.38.
7. $0.151.
8. 0.096\frac{1}{4}$.
9. $0.1065.
10. 0.160\frac{1}{4}$.
11. $0.2085.
12. $0.1345.

Lesson 65. Page 227.

1. $134.02.
2. $1940.
3. $54.79.
4. $128.17.
5. $28.94.
6. $97.06.
7. $36.68.
8. $26.93.
9. $61.87.
10. $146.
11. $160.65.
12. $133.88.
13. $104.22.
14. $119.82.
15. $530.25.

Lesson 66. Page 228.

1. $1.25; $248.75.
2. $14; $686.
3. $8.94; $966.06.
4. $3.54; $421.46.
5. $12.38; $1087.62.
6. $12; $1188.

Lesson 67. Page 229.

1. 5 days.
2. $720.
3. $342.25.
4. $0.30.
5. $300.
6. $22.19.
7. $12.55; $740.45.
8. $3.54; $446.46.
9. $913.67.
10. $474.60.
11. $224.81.
12. 12$\frac{1}{2}$%.
13. 3$\frac{1}{2}$%.
14. $3000.
15. $475.

MISCELLANEOUS PROBLEMS.

1. 1566.	40. 23,947.	79. 25.0826.	118. $73.74.
2. 1903.	41. $312.24.	80. 869.376.	119. $97.87.
3. 1482.	42. $279.84.	81. 488.5137.	120. $131.67.
4. 1804.	43. $282.44.	82. 82.1884.	121. $119.35.
5. 1200.	44. $320.89.	83. 121.9841.	122. 130.28.
6. 2196.	45. $267.46.	84. 633.7197.	123. 0.4808.
7. 1270.	46. $412.33.	85. 897.306.	124. 1.7134.
8. 1519.	47. $251.37.	86. 367.4876.	125. 3.5564.
9. 2081.	48. $310.35.	87. 258.3662.	126. 5.2714.
10. 1566.	49. $254.20.	88. 357.1753.	127. 3.5058.
11. 1837.	50. $283.01.	89. 596.	128. 4615.71.
12. 1596.	51. $292.16.	90. 2104.	129. $260.93.
13. 1458.	52. $257.08.	91. 1105.	130. 0.57135.
14. 1738.	53. $260.78.	92. 45,538.	131. 3.7886.
15. 1850.	54. $239.62.	93. 8420.	132. 4377.52.
16. 1218.	55. $262.39.	94. 62,876.	133. 262.309.
17. 2368.	56. 27,884.	95. 14,793.	134. 3.16363.
18. 1770.	57. 27,497.	96. 18,892.	135. 0.4234.
19. 1512.	58. 27,154.	97. 6578.	136. $14.18.
20. 1159.	59. 27,398.	98. 38,547.	137. 15.4776.
21. 1339.	60. 27,128.	99. 33,678.	138. 119.004.
22. 1635.	61. $1195.76.	100. 36,092.	139. 908.318.
23. 1706.	62. 3500.95.	101. 11,078.	140. 15.543.
24. 1982.	63. $5248.27.	102. 23,634.	141. 0.0018.
25. 1264.	64. $18,683.85.	103. 16,032.	142. 0.9412.
26. 1617.	65. 1,563,809.	104. 11,595.	143. 0.2312.
27. 2306.	66. 2,982,482.	105. 191.	144. 4.4907.
28. 2035.	67. 2,718,988.	106. 17,426.	145. 152.001.
29. 2004.	68. 3,224,531.	107. 28,156.	146. $51.77.
30. 1564.	69. 4,442,359.	108. 16,715.	147. 19.4132.
31. 22,025.	70. 2,332,108.	109. 48,060.	148. 6.3797.
32. 15,744.	71. 3.764,557.	110. 34,727.	149. 0.0575.
33. 15,804.	72. 4,155,684.	111. 7773.	150. 25.0727.
34. 24,424.	73. $11,056.41.	112. 53,745.	151. 49.8014.
35. 20,515.	74. $4282.44.	113. 15,129.	152. 0.172.
36. 17,455.	75. $12,111.50.	114. 28,218.	153. 13,824.
37. 23,384.	76. $6653.52.	115. 2126.5.	154. 23,886.
38. 24,598.	77. 529.162.	116. $39.28.	155. 198,432.
39. 23,661.	78. 30.6918.	117. $40.69.	156. 267,054.

157. 240,632.
158. 2122.88.
159. 59,895.
160. 146,316.
161. 344,396.
162. 651,714.
163. 4,056,840.
164. 9,799,576.
165. 16,538,112.
166. 15,374,074.
167. 18,060,048.
168. 11,055.8.
169. 17,707.3.
170. 31,390.
171. 90,713.6.
172. 502.469.
173. 0.33372.
174. 0.38465.
175. 1.37214.
176. 696.96.
177. 229.8912.
178. 34,385.8855.
179. 835.284.
180. 4690.476.
181. 4,703,699.
182. 1,869,120.
183. 215,152.
184. 216,750.
185. 2,972,000.
186. 2,685,600.
187. 3,760,380.
188. 2,471,841.
189. 4,128,168.
190. 0.365856.
191. 28.21.
192. 325.728.
193. 17.1854.
194. 37.5699.
195. 6257.44.
196. 5250.189.
197. 1834.3724.

198. 800,875.76.
199. 2856.75.
200. 2069.2.
201. 22,792.5.
202. 47,153.96.
203. 0.540465.
204. 12.66085.
205. 18.72.
206. 0.376516.
207. $1640.723.
208. $2533.864.
209. $283.7625.
210. $2730.285.
211. 3020.6016.
212. 55,632.3415.
213. 0.0269871.
214. 36,994,875.
215. 10,642.2336.
216. 0.48543716.
217. 144.
218. 204.3953.
219. 102.9691.
220. 96.2941.
221. 1135.6076.
222. 737.9512.
223. 1148.8684.
224. 1070.4286.
225. 1367.6290.
226. 676.0781.
227. 668.7129.
228. 291.3951.
229. 230.7139.
230. 53.6192.
231. 4415.6667.
232. 685.9157.
233. 1021.7616.
234. 937.7367.
235. 2740.1036.
236. 2438.4274.
237. 365.
238. 68.5237.

239. 64.5390.
240. 0.7209.
241. 13,100.
242. 0.005.
243. 96.
244. 73.1707.
245. 1.3305.
246. 246.4789.
247. 1496.3663.
248. 0.0904.
249. 22.8834.
250. 1.3267.
251. 0.1158.
252. 0.0195.
253. 0.3902.
254. 0.04.
255. 33.3333.
256. 0.1204.
257. 10.3311.
258. 1340.48.
259. 1540.
260. 1566.9867.
261. 500.
262. 0.2571.
263. 0.02063.
264. 0.0962.
265. 15.8556.
266. 16.
267. 2304.
268. 11,888.
269. 1825.
270. 13,100.
271. 3000.
272. 24.
273. 0.0134.
274. 101.16.
275. 11.3767.
276. 10.0150.
277. 30.7882.
278. 302.4037.
279. 75.0186.

280. 6.5455.
281. 17,948.
282. 308.53.
283. $728.60.
284. 251,964.
285. 236.08.
286. $333.50.
287. 87,827 sq. mi.
288. 1932.
289. $6805.65.
290. 623,800,367.
291. 4,700,745.
292. 1,387,568.
293. 67,384 sq. mi.
294. 2582.
295. $26,608.10.
296. $2886.76.
297. $2764.33.
298. $18,660.
299. 8302 lb.
300. 500.18 mi.
301. $273,484,572.
302. 15,894.
303. 96,970.
304. $3046.76.
305. 62.039.
306. $1,227,023,302.
307. $3590.
308. 876.
309. $290.52.
310. 146.2.
311. 2,321,644.
312. $153,978.82.
313. 1028.
314. 6543.20.
315. 69,686.
316. 72,447.
317. 17,099.611.
318. 2467.
319. 44.78.
320. 557.7.

321. $1543.75.
322. 12,466,467.
323. 428.
324. $2611.50.
325. $376.30.
326. $3846.75.
327. $5635.22.
328. $6436.50.
329. $15,592.60.
330. 8.153.
331. $2628.35.
332. 10.788.
333. $2585.25.
334. 284.
335. 15,990.
336. 73.78.
337. 540.49 mi.
338. 18.2.
339. 0.562.
340. 8.991.
341. 0.978.
342. 31,582.
343. 415,451.
344. 22,797.
345. 11,787.
346. 487,578.
347. 66,119.
348. $189.75.
349. 43.85.
350. $54,474.65.
351. 193.76 mi.
352. 408.
353. 174,149.
354. 179,358.
355. $2,340,675.
356. $1755.78.
357. 1357.
358. $1535.15.
359. 4720.
360. $6224.75.
361. $4522.77.

362. $837,864.
363. 139,806 ft.
364. 78,240.
365. 425.8536.
366. 528.769.
367. $427.50.
368. $780.
369. $3043.
370. $1099.
371. $756.
372. $675.
373. 32,256 ft.
374. $6.19.
375. $805,288.05
376. 56,940.
377. $19,443.05.
378. 195,216.
379. 2176.
380. $261.60.
381. 1564.
382. 1140 ; 68,400.
383. $207.36.
384. 24,993.75.
385. $308.47.
386. 1757.5.
387. 122,573.94.
388. $10,638.
389. 55.29216 in.
390. 8.59437 in.
391. $18.58.
392. $2.24.
393. $156.25.
394. 64.375 lb.
395. 91,698,000 mi.
396. 22,673.
397. 8772 lb.
398. $499.50.
399. $71.11.
400. $6,287,408.22.
401. $29,897,808.92.
402. $454.44.

403. 3648.

404. $56,546.10.

405. 2436.

406. 9119.25.

407. 503,194.

408. 84.8232 in.

409. 155.

410. 644,496.

411. 2,239,052.

412. $32.93.

413. $295.76.

414. $1406.25.

415. 2,985,984.

416. $107,564,745.75.

417. $8759.70

418. 24.

419. 5.82.

420. 6.2887.

421. 224.

422. 42 mi.

423. 78.

424. 23.

425. 2.8.

426. 362.

427. 364.

428. 80.

429. 18.

430. 1953.

431. 45.

432. $1.75.

433. 36 bu.

434. 18.

435. 27.

436. 280 mi.

437. 16.

438. $1.80.

439. 285 casks; 30 gal.

440. 33.

441. 19.

442. $5.

443. 104.

444. 1213.

445. 83,521.

446. 29.7415 in.

447. $40.

448. 14.

449. 34.65925.

450. $8.75.

451. 48.

452. 62.

453. 528.

454. 24.

455. 385.4167.

456. $136.99.

457. 19.

458. 28.

459. 53.1905.

460. $44,315.76.

461. 55 ; 32.

462. 150.

463. 60.

464. 7.0076.

465. 800.

466. 29 bu.

467. 32 bu.; $0.20.

468. 82.

469. 96.

470. 118.

471. 33,696.

472. 3 A. 132 sq. rd. 21 sq. yd.

473. 4 mi. 3360 ft.

474. 521.

475. 67 yr. 9 mo. 22 dy.

476. May 5, 1821.

477. 60 mi.

478. 13 A. 119 sq. rd. 28 sq. yd. 2 sq. ft. 36 sq. in.

479. $90^{-}\frac{1}{2}$.

480. $682.88.

481. $3.24.

482. 365 dy. 6 hr. 9 min. 12.96 sec.

483. 29 dy. 12 hr. 43 min. 12 sec.

484. $\frac{1}{7}$.

485. 27.3215 dy.

486. 285.

487. $200.81.

488. 35 lb. 11 oz.

489. 2160.

490. 1687.5.

491. 14.56 qt.

492. 800.

493. 81 gal. 3 qt. 1 pt.

494. 2400.

495. 72,000.

496. $29.71.

497. $22.35.

498. $127.25.

499. $238.70.

500. $51.49.

501. $43.30.

502. 7 ft. 4 in.

503. 130.

504. 58,500 sq. yd.

505. 2304.

506. 756 cu. ft.

507. 154 sq. in.

508. 9482 ft.

509. 25 rd.

510. 13 ft.

511. $3\frac{9}{11}$ ft.

512. 14,300.

513. $25.67.

514. 15.

515. $21.09.

516. 9.

517. 30 yd.; $31\frac{1}{2}$ yd.

518. $104.86.

519. $57.17.

520. 6.

521. $17.50.

522. $22.

523. 12.

524. 54.

525. 1800.

526. 162.

527. $262\frac{1}{2}$.

528. 144.

529. 80.

530. $26\frac{5}{12}$.

531. 15.

532. $167.19.

533. 6 hr.

534. $\frac{23}{48}$.

535. 0.6; 0.875; 0.5625; 0.3333; 0.2917; 0.5265.

536. $\frac{7}{8}$; $\frac{13}{20}$; $\frac{77}{200}$; $\frac{11}{25}$.

537. $21\frac{3}{4}$.

538. $\frac{10}{63}$.

539. $21\frac{7}{16}$.

540. $157.50.

541. $1\frac{5}{18}$; $\frac{7}{18}$.

542. $\frac{1}{5}$.

543. $21\frac{2}{9}$.

544. $3\frac{1}{5}$.

545. $2\frac{3}{23}$.

546. $92\frac{7}{20}$.

547. $3\frac{1}{2}$.

548. $\frac{9}{80}$ increased.

549. Horse, $350; carriage, $100.

550. 100.

551. 1120.

552. $14\frac{1}{7}$.

553. $1.62.

554. $39.63.

555. $4.65.

556. 24.

557. $\frac{31}{45}$.

558. $\frac{1}{5}$; $\frac{1}{4}$; $\frac{1}{3}$; $\frac{7}{8}$; $\frac{5}{6}$; $\frac{9}{20}$; $\frac{1}{16}$.

559. 50%; $33\frac{1}{3}$%; 75%; $83\frac{1}{3}$%; 60%; 90%; 65%.

560. 348.

561. $817.53.

562. Saltpetre, 1500 lb.; sulphur, 200 lb.; charcoal, 300 lb.

563. 9%.

564. 363.

565. 20%.

566. $16\frac{2}{3}$%.

567. 24.86%.

568. 180.

569. 6052.

570. $1\frac{1}{2}$%.

571. $3750.

572. $2500.

573. $13.05.

574. $598.74.

575. $11.25.

576. $1028.75.

577. $3619.39.

578. $1017.76.

579. $1231.51.

580. $298.47.

581. $174.98.

582. $222.25.

583. $198.49.

584. $230.52.

585. $31.25; $1218.75.

586. $34.21; $2190.79.

587. $8.16; $781.84.